INTEGRATED PRODUCT TESTING
AND EVALUATION

QUALITY AND RELIABILITY

A Series Edited by

Edward G. Schilling

Center for Quality and Applied Statistics
Rochester Institute of Technology
Rochester, New York

Additional volumes in preparation

INTEGRATED PRODUCT TESTING AND EVALUATION
A Systems Approach to Improve Reliability and Quality

REVISED EDITION

HAROLD L. GILMORE

Department of Management
Pennsylvania State University
Philadelphia, Pennsylvania

HERBERT C. SCHWARTZ

Senior Staff Consultant
AVCO Corporation
Woburn, Massachusetts

MARCEL DEKKER, INC.
ASQC QUALITY PRESS

New York and Basel
Milwaukee

Library of Congress Cataloging-in-Publication Data

Gilmore, Harold L.
    Integrated product testing and evaluation.

    (Quality and reliability ; 6)
    Includes index.
    1. Testing. 2. Reliability (Engineering)
3. Engineering inspection. I. Schwartz, Herbert C.
II. Title. III. Series: Quality and reliability ;
vol. 6.
TA410.G518   1985   620'00452   85-20445
ISBN 0-8247-7470-1

MARCEL DEKKER, INC.
270 Madison Avenue, New York, New York 10016

ASQC
230 West Wells Street
Milwaukee, Wisconsin 53203

Current printing (last digit):
10 9 8 7 6 5 4 3 2 1

PRINTED IN THE UNITED STATES OF AMERICA

Dedicated to our families who, in turn, are dedicated to us

Foreword

It is said, that the ancient Greek philosopher Thales was once walking along a road by night with his head up, studying the stars. Rapt in concentration, he stumbled and fell into a ditch and was forced to yell for help. An old woman, passing by, laughed scornfully and said, "Here is a man so busy studying the vast distances above his head that he cannot see what is under his very feet."

All science is, in a way, in the position of old Thales. It is performing miracles undreamed of only ten years ago. It has worked out the biological dictionary that translates the structure of deoxyribonucleic acid (DNA) into the structure of the protein molecule. It has built lasers that shoot out streams of light with properties never seen in nature and which are capable of making three-dimensional holographs. It is transplanting hearts and contemplating transplanting brains. It drops instrument packets through the atmosphere of Venus and makes ready to land men on Mars.

Yet while we soar into the scientific heavens, we stumble over our commonplace technological feet. The products that we use every day come to us with built-in obsolescence that sometimes strikes the moment we open the package. Each new convenience comes wrapped with its full quota of inconveniences. New automobiles age before our eyes, toasters scorch neatly through their heating elements, batteries fizzle, and the consumer burns. Is this a conspiracy? Are industrialists stockholders in aspirin-manufacturing firms?

Not at all. It's just the competition between the glamorous and the prosaic, between the stars overhead and the ditch underfoot. In being forced to choose between a career devoted to working out the theory of sophisticated space-vehicle components and one devoted to working out prosaic testing procedures involved in making flashlight switches, the stars win—and we fall into the ditch.

This must not continue. It is the purpose of this volume to help make sure that it does not continue. With improved product testing and greater care in manufacturing, it may just be that when our astronauts land on some distant planet with perfectly functioning retro-rocket procedures the switch on the dollar flashlight they bring along won't jam.

ISAAC ASIMOV

Introduction

The high quality of the products that pour from the cornucopia of today's science and technology results from the creation of an entirely new function: test and evaluation engineering.

Product testing, which not too long ago was merely an afterthought of the procurement process, has now become an integral part of the entire research and development cycle. It is no longer performed to find out what has been built. Now the design incorporates data from the test and evaluation function.

In the missile field, 25 years ago, it was customary for one group of people (consisting of scientists, engineers, and technicians) to perform design, development, and testing, and for another group (consisting of engineers, technicians, and craftsmen) to take care of the manufacturing. When the item was completed, it went back to the first group for testing and evaluation. Problems were found in workmanship, in parts and components, and in subsystems and systems that did not perform as intended. It was clear that much more thorough surveillance of the entire process was mandatory.

Over the years, many techniques have been developed for in-process inspection, review, and testing. Experience proved that most spacecraft and missile failures did not result from the failure of components (because of their complexity), but from the poor quality of parts that make up the components. A system of pyramidal testing evolved in which we first tested the parts, then the components, then the subsystems, the systems, the stages, and eventually the entire launch vehicle before it was committed to the firing pad.

Feedback to the manufacturer and the designer was established and progress multiplied rapidly, for achievement of reliability goals must start with the design.

Ultimately we were able to furnish detailed requirements and procedures to our contractors for quality assurance and reliability. Special care was taken to combine the paper effort and the hardware effort into an integrated whole.

The successes that the National Aeronautics and Space Administration has enjoyed with the Saturn launch vehicles can be attributed to increasing use of these principles which have evolved over the years.

Organizationally, a clearly identified group must be given the responsibility for the reliability function, and its head must have unimpeded access to top management in order to have the necessary authority for a proper discharge of his or her responsibilities across the board.

Persons skilled in the new role of product test and evaluation will become increasingly valuable to management in the future as technologies expand, as missions become more difficult, and the requirements for hardware performance become more and more stringent.

WERNHER VON BRAUN
1912-1977
National Aeronautics and
Space Administration
Marshall Space Flight
Center, Alabama

Preface to the Revised Edition

In the years since this book was originally published, several forces have converged to focus increased attention on product testing and evaluation. Three of the most significant include technological change, government regulations, and world class competition in the sphere of product performance and quality.

Technology in product design and manufacture is subject to an accelerating rate of change, especially in the electronics industry. Technological innovation and sophistication has resulted in today's customer expecting more capability for his money, along with greater reliability, safety, and serviceability. Satisfaction of these expectations requires the application and interaction of key technologies and principles: computer-aided design and computer-aided manufacture (CAD/CAM), producibility, automation, and one introduced in this book, computer-aided product evaluation (CAPE).

Government has increasingly expanded its sphere of influence over the design, manufacture, and sale of products. In 1972 Congress created the Consumer Product Safety Commission (CPSC) in an attempt "to protect the public against the unreasonable risks of injuries and deaths associated with consumer products." Every year, however, consumer products are involved in the deaths of an estimated 28,000 Americans and in injuries to 33 million others.

Product recall is the most obvious evidence of the success of CPSC activities in identifying product hazards and getting them out of consumer homes and off retail shelves. Some of the most noteworthy include the Firestone 500 radials; botulism outbreaks from Bon Vivant vichyssoise; Eli Lilly's Oraflex; and Johnson and Johnson's—through no fault of its own—Tylenol, a successful and healthy product. On the other hand, where there are product failures it reflects adversely on the product testing and evaluation system.

Finally, the world market is filled with a staggering number and variety of products requiring evaluation to see if they are viable, safe, and reliable. A recent study by Booz, Allen & Hamilton shows that more new products failed in 1980 than did in 1968—65% successful in 1980, 67% successful in 1969. One must conclude that product evaluations have not improved over the ensuing 13 years.

We know that there is a continuing need for this book; we hope that if put to use it will contribute to an improvement in the ability of American manufacturers to design, manufacture, and sell products inferior to none available anywhere in the world.

Harold L. Gilmore
Hershey, Pennsylvania
Herbert C. Schwartz
Woburn, Massachusetts

Preface to the First Edition

This book presents an economical and proved methodology of integrated product testing and evaluation (PT&E) as it is performed in industry today. Product testing and evaluation is not a new field, but recent stress on product quality and reliability has caused significant changes in its approach and execution. Moreover, the limited current literature applies only to specific areas of testing and therefore is of restricted use and interest. Consequently, the extent of PT&E as a comprehensive working tool is being offered for the first time.

We have attempted to resolve the technological gap that has existed by presenting fundamental and advanced concepts, principles, and considerations of the PT&E function, including such specific topics as organization planning, test and statistical methods, data-reporting systems, and environments and their control. How the reader chooses to formulate and adapt these methods will depend on his or her awareness of existing practice, on the type of equipment concerned, and also on the specific problems that he or she wants resolved.

The comprehensiveness of this volume will appeal to technical management because it is the attitude of technical management that establishes the extent and effectiveness of product evaluation activities throughout the organization. Technical management must realize the need for PT&E and the contribution it can make to successful business operations.

The most important audience for this book will consist of design, system, development, test, reliability, manufacturing and quality control engineers who are involved in the detailed activities that we describe. The knowledge gleaned from this study will enhance the value of these engineers to their employers. A unique feature is its coverage of industrial, aerospace, and commercial applications. Numerous examples are given from our management and technical experience in defense, commercial, and industrial organizations to ensure authoritative case histories. Significant consideration has been given to the technical utilization aspects of the various programs.

To all those involved either directly or indirectly in product testing and evaluation, we believe this book will play a major part in their everyday thinking and practices, thereby leading to improved quality and reliability of product.

HAROLD L. GILMORE
Syracuse, New York
HERBERT C. SCHWARTZ
Peabody, Massachusetts

Acknowledgments

It is with pleasure that we acknowledge the many sources noted throughout the book from which materials for figures, tables, and text were obtained. Our search for relevant materials was met with enthusiastic support from individuals, industries, organizations, and institutions alike. In particular, we want to express our appreciation to all those who provided material which has been updated and retained from the original edition, Laura J. Turbini of AT&T Technologies (and its publication *The Engineer*), and Mr. Bill Parker and his associates of Boeing Aerospace Company.

We also wish to recognize E. J. Carrubba of the Digital Equipment Corp., Arthur Durling of the AVCO Corp., and D. R. Earles of the Raytheon Corp. for reviewing the revised edition and providing valuable suggestions and encouragement.

Our continued appreciation to Isaac Asimov and the late Wernher·von Braun for taking time from their busy schedules to read our original manuscript and for writing the Foreword and Introduction.

We also wish to thank Janie Fuller for her unselfish and untiring effort in typing this revision.

Contents

Chapter 1

A Systems Concept: The Economics of Product Evaluation

The explosive advance of industrial, military, and aerospace technologies has given rise to an entirely new scientific function—that of test and evaluation engineering. What once, not too long ago, was an afterthought of general operations has now matured into an integral part of the product development process. Product testing and evaluation (PT&E) is no longer performed to find out what has been designed but, instead, the design is based on data systematically derived from the test and evaluation function.

At the outset, let us define where we are going and what we hope to accomplish. The main thesis is that the product test and evaluation function is an autonomous activity comprising a vital link in the product development cycle and therefore requires the application of concepts that are commensurate with its role. We feel that reading this book and, in turn, gaining an appreciation of the scope and content of the product evaluation task will allay many fears that this work is a tribute to the function written by two of its devotees.

Furthermore, we have not approached the task as "engineers" or as "managers," since our purpose was to bridge the gap between them. Our aims are twofold: (1) to show management the value of the function, and (2) to show the engineer the need for applying management-oriented techniques in decision making and program formulation. Therefore, this book includes a blend of both management and technical considerations, and it is not expected that all will be of equal value to each reader. Since individual growth is normally upward, the technical material will have immediate usefulness to engineers, and the management material will provide them an insight into the supervisory parameters of PT&E.

The basic plan of the book is indicated by the titles of the chapters. Our first chapters cover test and evaluation on an integrated, comprehensive system basis, relying primarily on the use of examples to demonstrate the principles involved. They also deal with managerial aspects and illuminate some of the major areas with which the executive must cope in order to fulfill this function successfully. The reader is left to his or her own application decisions, inasmuch as the degree of detail presented may or may not be necessary, depending on the particular situation. Hopefully the examples presented will give the practitioner the framework upon which to develop the program most appropriate for his or her purposes. Ensuing chapters are devoted to actual test and evaluation technical considerations. We have presented specific tests and procedures, statistical considerations, and product life evaluation techniques.

The concluding chapters of the book present sources of test information and examine the PT&E transition.

Throughout the book we have stressed that knowing when to employ the techniques of test and evaluation requires a broad-based understanding of methods and approaches. We have attempted to provide the base and, at the same time, sufficient background to interest the practitioner. The book will serve as a day-to-day tool for some applications and a reference work for others. It will be used to:

- Help develop a test program tailored to a specific product or a specific contract.
- Serve as a guide to a company in its regular test work.
- Help companies establish a testing program or organize one.
- Train new engineers entering the field.
- Serve as a college text for schools that are developing quality control curricula.
- Serve as a guide to companies that have little testing activity, but who contract out their testing.
- Outline the responsibility for evaluating vendors' testing programs and certifying vendors.

As an economic tool, the book will help to prevent overdesigning a product, yet make the product as reliable as it needs to be. Also, as an economic tool, it could help the company determine warranty and guarantee provisions and forecast their cost. As a marketing tool, PT&E results can prove company competence in helping to obtain contracts, and they can be used to advertise the merits of the product. For planning, PT&E results can help to make the decision of when to manufacture a new product. As a technical tool, it can help to solve design problems such as selection of materials, and it can help to evaluate the product at various design stages. In a legal sense, PT&E results can prove that products meet contract provisions and legal requirements.

The establishment of a PT&E organization equipped with the latest test equipment and staffed with competent engineers is no assurance that an acceptable contribution will be realized. Testing for the sake of testing is of no value. Testing is a service and, as such, must serve a useful purpose to be worth performing. Therefore, the PT&E function must be related to the functions that can act on the results and recommendations derived from it.

To provide the reader with a systems perspective of product evaluation, this chapter presents the results of an investigation into conformance quality control and its cost as practiced by manufacturers of consumer products. Quality control costs were selected because they offered the unique opportunity to integrate the systems concept with a specific aspect of product evaluation. The underlying concept of resource allocation is relevant to the other areas of product evaluation and could have been applied to any one of them. However, since quality cost data exists in limited quantity, the opportunity to incorporate it here was taken.

Commencing with a discussion of systems concepts and work sampling and their relationship to product conformance quality control the economic theory underlying quality costs and their allocation is developed. Quality cost data analysis and presentation provides an opportunity to compare company practice with theory and to assess the extent to which economic theory is put into practice.

SYSTEMS CONCEPT

This discussion begins with a systems overview and proceeds to an exploration of the firm as a system consisting of several subsystems. The quality control subsystem is the subsystem which is studied in detail.

OVERVIEW

Systems concepts have become key considerations in the study of the organization and management of both public and private institutions. A systems philosophy is a way of thinking about complex human activity. It facilitates recognition and understanding of interrelationships among the various activities that are involved in any modern organization. However, each manager must also take a contingency view as systems concepts are employed in the diagnosis of specific situations. Such a view will facilitate the development of managerial action that is appropriate in the specific circumstances.

This discussion involves the application of systems concepts to the specific area of product conformance quality control. The theoretical aspects of quality control economics applied to specific industry circumstances is examined to identify management behavior to achieve product quality objectives.

Kast and Rosenzweig[1] define a system as "an organized or complex whole, an assemblage or combination of things or parts forming a complex or unitary whole." They go on to point out that there are hierarchies of systems; a typical system will include subsystems and at the same time be part of a suprasystem. While the parts of a system are important considerations, systems theory concentrates on interaction, interrelationships, and integration of parts or subsystems into a whole.

Boulding's[2] hierarchy of systems encompasses a range which includes stable structures such as a building, to transcendental systems such as philosophy or religion. Here we apply the systems concept to the second highest order in Boulding's hierarchy, that being a social organization. We are also dealing with decision processes and behavioral responses to organizational control techniques.

Other properties of organizational systems are worth mentioning to avoid leaving the notion that the foregoing comments are exhaustive. Listed below are several widely accepted properties attributed to organizational systems.

- Contrived
- Boundaries
- Hierarchy
- Negative Entropy
- Steady State or
  Dynamic Equilibrium

- Feedback
- Adaptive and
  Maintenance Mechanisms
- Growth via
  Internal Elaboration
- Equifinality

It is not necessary to go into a detailed explanation of the above to meet our needs. It is sufficient to note that a complete understanding of "systems concepts" requires familiarity with all of the above. Any study of a social organization will provide ample evidence of these properties. If aware of them at the outset, the analyst will be immediately conscious of their presence.

MANUFACTURING SYSTEM

Manufacturing organizations are of primary consideration and may be depicted in accordance with the systems concept as illustrated in Figure 1.

The subsystems consist of production, finance, marketing, and personnel. Depending upon the specific company, there may be more or less significance to one than the other. However, the managerial subsystem must always be present to integrate actions to achieve effective and efficient performance of the whole.

PRODUCTION SUBSYSTEM

The production subsystem also consists of several subsystems. The lower the subsystem in the hierarchy, the more "task" oriented the subsystem be-

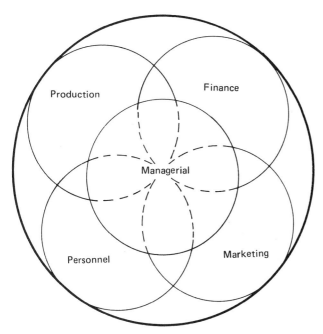

Figure 1 Typical manufacturing organization.

comes. For example, a typical production subsystem is depicted in Figure 2. Note the tasks that are represented and their planning and control orientation.

As in the manufacturing system diagram, an integrating subsystem is shown here also—management information. In this age of computerized data gathering, analysis and dissemination, the management information system serves as a driving force for planning and control activities. The information system provides both the collection and dissemination of information for managerial decision making. In some instances, the control system is embedded in the management information system itself just as the computer is embedded in the production equipment. Product quality can be controlled continuously by real-time, on-line adjustments to production processes. It is the quality control subsystem to which we now turn.

QUALITY CONTROL SUBSYSTEM

Quality control is the function of assuring that the attributes of the product conform to prescribed standards, and that their relationships to one another

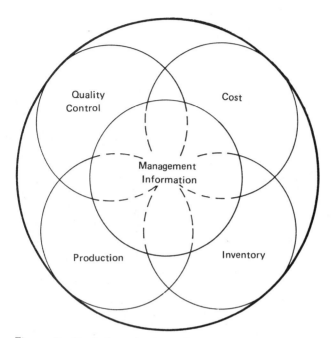

Figure 2   Typical production subsystem.

are maintained.[3] It involves the maintenance of dimensions and other product
characteristics within a predetermined specific plan—a responsibility which
is substantially broader than rejecting unsatisfactory parts.

A properly conceived quality control program will make significant savings
in both direct and indirect costs. For example, effective control of the quality
of raw materials will prevent the processing of nonconforming materials.
Production time, which has been spent in producing or reworking defective
parts, now can be used to increase the quantity of production. A good product
builds customer goodwill. Workers take pride in producing quality products
or in being associated with a company that does.

Although product quality depends on many factors, the standards for it
are determined by the customer or user. For certain products, quality standards
are dictated by the government, i.e., pure food and drug regulations.
Nevertheless, in one or more of these ways, taken separately or in combination,
the product and what is to be controlled is established.

All control systems have several elements consisting of the following: (1)
a standard, (2) performance monitor, (3) a comparator, (4) an effector, and
(5) a feedback loop through which corrective action may be taken if
appropriate. In the usual situation, the inspection group serves as the com-

parator in the quality control system. This group determines what to examine, when, and how much. It is a major economic decision to know where the optimum inspection level lies between 100% examination to minimum sized samples or no inspection at all.

Many factors enter into this determination and it is this managerial decision which is of interest to us. Our objective is to determine to what extent work sampling resource allocation decisions are in conformance with theory, i.e. in line with the optimum allocation of prevention, appraisal, and failure-related quality activity as depicted in Figure 3.

WORK SAMPLING

Lazzaro[4] notes that ". . . about fifteen years ago the technique (of work) developed and used in the plant was successfully applied to office operations." Langevin[5] prepared a management briefing on the subject about that time. In the interim many additional applications have been made of work sampling to office and service industry applications. However, the development of quality cost identification, collection, and analysis within the service industry has not followed as rapidly. This simply reflects the maturation of the field and agrees with the development of quality control and cost systems within

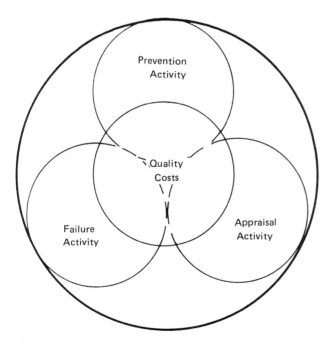

Figure 3 Quality control subsystem.

industry. Even now quality cost collection is not a uniform practice within industry. Any study of this practice will quickly identify the nonavailability of quality cost data although the situation is better today than it was ten years ago. Today, because of competitive reasons, there appears great reluctance to release the cost information when it is available. Perhaps in another five years a study of office and service industry work sampling costs will be possible.

ECONOMICS OF QUALITY CONTROL

Unless one is familiar with the specialized area of quality control, selected terms must first be defined. Failure to do so will create confusion and render the ensuing discussion concerning quality cost concepts and their allocation meaningless.

TERMINOLOGY DEFINED

The meanings of three terms are essential, not only because of their significance but also because of their vagueness and variety of use. The three terms are "quality," "control," and "quality control."

Quality. In layman's language quality means general excellence and when applied to consumer goods implies excellence in the properties looked for and expected in the product; for example, appearance, ruggedness, taste, performance, and potency.

A related definition states that "quality is the degree to which a *specific* product satisfies the wants of a *specific* consumer."[6] Whether the product possesses these properties depends on whether they were designed into the product, and if so, the extent to which the manufactured product conforms with the established specification.

Still another definition states "quality is the degree to which a specific product *conforms* to a design or specification."[7] This is known as conformance quality and the definition applied here.

Control. Just as "quality" has a number of definitions, "control" also has a variety of meanings. As traditionally conceived, control means making sure that actual results conform to desired results, and this involves three basic functions: (a) setting standards of satisfactory performance, (b) checking results to see how they compare with the standards, and (c) taking corrective action where actual results do not meet the standards.

The notion of "control" is used here in conjunction with the term quality. Therefore, it is defined as "the action taken which adjusts operational variables to achieve predetermined standards of quality." This introduces the third term "quality control."

Quality Control. The application of this term within industry is by no means uniform, leading to ambiguity and confusion as to just what is meant.

In some companies quality control is broadly used to connote the functions and/or aggregate of duties which must be performed in order to carry out quality objectives. In other companies quality control is narrowly applied and may refer to a specific activity such as inspection, test or statistical analysis. When "quality control" is used to describe a specific organizational unit the tasks performed may lie anywhere along the continuum of applications described above. "Quality control" is: "all of the activities which must be carried out to insure that the specified quality level of the product is achieved and maintained."

INDIVIDUAL MANUFACTURER PERSPECTIVE TOWARD QUALITY

There are a number of views toward product quality among which are those of the economist, government, consumer groups, industry, and the individual manufacturer. The individual manufacturer's view is of particular importance. His view is mainly one of achieving customer satisfaction and is directly related to the market niche that the company wishes to serve. The manufacturer takes those actions considered necessary to insure that customer expectations are satisfied. Through market research, management strives to determine specific consumer wants, specifications of competitive products and marketplace quality and price structure. Having determined this information the inherent quality of the product design can be established in the form of technical specifications. In light of the above, the manufacturer's view is clearly toward the customers and competitors. The development of technical specifications, however, requires that the production process also be considered.

Even where product technical specifications are properly established for the desired market, production processes may or may not operate properly. Improper operation will effect production quality by causing a product to be improperly manufactured and/or shipped only to fail in use. This is production quality and is a difficult, costly control task. In carrying out this task the manufacturer is concerned with those product characteristics that are critical from a control point of view. Therefore, the operations producing these characteristics receive close attention. It is interesting to note that these control characteristics may or may not be important from the consumer (the customer's) point of view.

Nevertheless, a product produced in conformance to specifications and within the control limits of the production process will satisfy the quality requirements of the consumer if they have been properly defined by the manufacturer. It is the economics of this production-related quality activity, heretofore defined as conformance quality control, and its cost that is discussed below.

Conformance Quality Control Cost. The background information provided thus far has defined certain key terms and provided conceptual clarification.

There remains one final point of clarification—*conformance quality control cost.*

It is prudent to note that conformance quality control cost is only a portion of total product quality cost. This total cost of product quality is made up of the costs associated with inherent design quality as well as the costs associated with conformance quality control. The nature of conformance quality control will be discussed to delineate the difference between design and conformance quality and in turn, to explain the conformance concept.

Product conformance quality control is concerned with obtaining conformance to established design quality as reflected in product specifications. In general terms conformance quality control involves the task of: (1) selecting and establishing appropriate standards, processes and procedures (prevention activity), (2) monitoring performance to established standards (appraisal activity), and (3) taking corrective action where necessary to bring operations and/or the product into line with established standards (corrective activity). The total cost of conformance quality control results from the interrelationship of the separate costs associated with the implementation of these tasks. To explain the concept consider this illustration.

Product conformance to specification is the objective of quality control. Figure 4[8] depicts this situation in terms of process cost only. Note that as the process cost to achieve increasing product conformance rises at an increasing rate, concurrently the cost due to specification deviations decreases but at a lower rate. The quality control task is to achieve conformance quality where total conformance costs are minimum. This establishes the optimum process cost and specification deviation cost.

Conformance quality control implementation consists of prevention, appraisal and failure activities, each of which has costs. Masser[9] classified quality costs in terms of prevention, appraisal, and failure activities.

> 'Prevention Costs' which are spent for the purpose of keeping defects from occurring in the first place. Included here are quality control engineering; quality planning; design of measuring equipment; quality training; such maintenance of tools as is essential to maintenance of product quality.
>
> 'Appraisal Costs' which include the expense for maintaining company quality levels by means of formal evaluations of product quality. Specifically included are incoming, process and final inspection and test, whether performed by production, inspection, or other means, maintenance of test and inspection equipment; set-up costs for inspection and test; cost of product destroyed, materials consumed and services used (power, steam, etc.) during inspection and test; quality audits; outside endorsements.
>
> 'Failure Costs' which are caused by defective materials and products that do not meet company quality specifications. Included are scrap; rework;

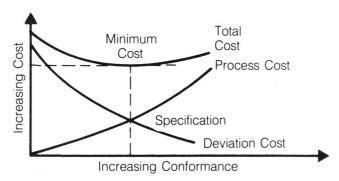

Figure 4 Process conformance quality cost.

cost of vendor relations on defective goods; cost of engineering work on factory defect problems; all cost of complaint adjustment; all costs for customer service due to defects.[10]

In order to measure the cost in each of the three activities of conformance quality control requires the identification of detailed work elements involved in each. Warren Purcell[11] and Armand Feigenbaum[12] have defined categories of work elements of which all or some are applicable to a particular company. Referring to the contributions of Feigenbaum, Masser, and Purcell, work activity descriptions were synthesized for use as models for the collection of the quality cost data presented later.

Allocation Theoretical Foundation. The theory concerning allocation of conformance quality control cost represents the application of microeconomic concepts to conformance quality control. The specific microeconomic concepts employed are "cost analysis" and "profit maximization."

The economic model for the allocation of quality assumes that the objective of the quality control manager is to achieve conformance quality control (at any given quality level) at minimum conformance quality control cost. This objective is to the quality control manager, the equivalent of profit maximization.

Second, conformance quality control is achieved through the simultaneous application of "prevention," "appraisal," and "failure" activities. The costs of prevention and appraisal are associated with efforts to assure production of perfect goods in the first place and failure costs are associated with having produced defective (nonconforming) goods.

Finally, the cost of prevention, appraisal, and failure activities are interrelated. Prevention and appraisal activities have as their objective the elimination of failure-related activities. Prevention activities are also oriented toward minimizing the need for appraisal (inspection and test) activities. The resultant cost relationship due to the process of minimizing the sum of these three related costs is examined. More specifically, the reported conformance quality control expenditure practice of companies is compared to the theoretically derived quality control cost relationships when the manufacturing plant is operated so that total conformance quality control cost is a minimum.

The theoretical relationship of the constituent costs of conformance quality control (prevention, appraisal, failure) rests on a logical and intuitive base. Presently there does not exist a rigorous justification for the exact form of the cost curves upon which to rely.

It is generally believed that the emphasis in achieving product quality control should be on prevention activities rather than appraisal or corrective (failure) activities. The logic behind this idea is that it is less costly to prevent defective products from occurring in the first place than allowing defects to occur and "after-the-fact" attempts to correct them. Embodied in this idea is the realization that there are more to failure costs than the tangible costs associated with warranty, scrap, and rework. Also included are intangible costs such as lost sales, lost goodwill, and added productive commitment required to make necessary product replacements and corrections.

Figures 5 and 6 depict, for the purpose of analysis, nonparametric quality cost models based on the foregoing concepts. They depict the interrelated constituent costs of conformance quality control as a function of prevention, appraisal, and failure activity.

Figure 5 depicts prevention and appraisal cost behavior as a function of prevention activity. Prevention costs are shown increasing linearly as prevention activity is increased.[13] The appraisal cost curve illustrates that as prevention activity is increased, the cost of appraisal can be expected to decrease but at a decreasing rate.[14] This curve approaches zero as prevention activity increases which reflects the reality that the complete elimination of appraisal activities (costs) is highly unlikely. There will always be some measurement required to determine the extent to which the manufactured product conforms to specifications. In addition, this curve indicates that some finite value of appraisal costs exists even when little or no prevention activity is performed. In that situation the manufacturer would be engaged in determining what the quality level of the product output actually was.

The sum of the prevention and appraisal cost curves yields a curve depicting the cost of assuring the desired quality is initially produced in the manufacturing operations.

Figure 6 illustrates the total cost of conformance quality control as a function of prevention and appraisal activity. The prevention and appraisal cost curve as depicted in Figure 5 is duplicated in Figure 6, along with the resulting expected relationship to failure cost. Failure cost is shown decreasing and

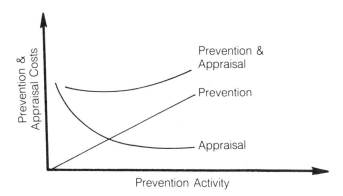

Figure 5 Quality assurance cost model (quality level fixed).

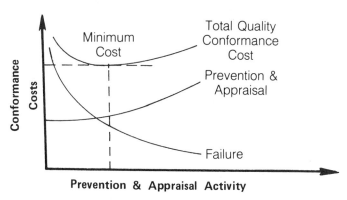

Figure 6 Conformance quality control cost model (quality level fixed).

approaching zero as prevention and appraisal activities are increased. The failure curve is shown approaching zero which reflects the idea that the complete elimination of product failure is an unlikely reality. There is bound to be scrap, rework, or customer service required in any manufacturing operation for a variety of personnel- and material-related causes. The curve also indicates that some finite value of failure cost exists when little or no quality assurance (prevention or appraisal) activity is performed.

The total conformance quality control cost curve is the sum of the two cost curves. It takes the familiar U-shape of the cost curves depicted by economic theorists—technically, a quadratic cost function.[15] The bottom point of the curve indicates the minimum cost combination of prevention, appraisal, and failure activity. It may be necessary for a manufacturer to produce to a quality cost level higher than the minimum cost point due to customer requirements. However, a manufacturer seeking to maximize profits would not be justified in operating at any lower prevention and appraisal level. To do so would result in larger costs than necessary at a lower level of conformance quality (higher failure costs).

Maximizing profits from a conformance quality control point of view is to minimize total conformance quality control cost. According to Figure 6 this occurs approximately where the prevention, appraisal, and failure cost curves intersect.[16] A manufacturing plant operating at minimum total conformance quality control cost should employ the indicated mix of prevention, appraisal, and failure activity. In terms of minimum conformance quality control cost, the combined cost of prevention and appraisal should approximately equal the cost of failure.[17,18]

The practice of industry is evaluated with respect to this allocation concept utilizing actual cost data collected in a survey.

QUALITY COST SURVEY DATA

Quality does not cost, it is free.[19] The debate among the proponents of each view goes on and on. Regardless of the view, however, resources are expended in the pursuit of product quality in one way or another in every organization. Presented below are the results of a recent survey into the quality control cost experience of a sample of U.S. manufacturing firms.

Some time ago a similar study was conducted.[20] A decision was made to revisit the quality cost scene to see what change, if any, could be detected in the expenditure experience of the same or similar corporations. Since the product cost experience of manufacturers is considered a measurable reflection of corporate quality control behavior, any change in managerial perspective and decision making toward quality should be evident in expenditure data. Leonard and Sasser [21] recently reported finding a shift, at the general management level, from an inspection-oriented, manufacturing-focused approach toward quality to a defect-prevention and company-focused strategy. A second important change they noted was in the view held by management of quality personnel. They were seen as managerially focused, planning and prevention oriented, assertive, powerful, responsible for preventing failures, and well respected. This was in sharp contrast with earlier views where quality personnel were viewed as technically and problem oriented, defensive, powerless, responsible for inspection and the "fixing" of failures, and not well respected. These changes should be reflected in the allocation and expenditure of corporate resources devoted to the quality control task. The research data bearing on this issue and the evaluation are reported below.

INDUSTRIAL ORIENTATION

First, however, is a brief description of the participating organizations. Names are not reported to protect confidentiality of the data and their source. Instead, Standard Industrial Classification (SIC) code numbers are used, as indicated in Table 1, to identify the respondents.

Another distinguishing feature of the study participants is their size in terms of employment level. Table 2 provides their distribution according to the number of people employed.

OPERATIONAL CLIMATE

Sales, profits, and output reported by the majority were satisfactory to good. Few considered these characteristics as unsatisfactory. All reported the product price trend was up over recent years and all attached major importance to quality in the sale of their products.

BUDGETARY RESPONSIBILITY

The survey results clearly indicated that quality control management established the quality budget which was, in turn, approved by top management. The

Table 1　Survey Respondents ($n = 17$)

SIC		Number of business units
20	Food and kindred products	2
22	Textile mill products	1
25	Furniture and fixtures	1
26	Paper and allied products	3
28	Chemicals and allied products	3
34	Fabricated metal products, except ordinance machinery and transportation equipment	2
35	Machinery, except electrical	1
36	Electrical machinery, equipment and supplies	2
38	Measuring, analyzing and controlling instruments, photographic, medical and optimal goods, watches and clocks	1
39	Miscellaneous manufacturing industries	1

Table 2 Company Size
$(n = 12)$

Number of employees	%
< 200	17
201 – 500	25
501 – 2000	17
> 2000	42

basis for the budget, however, was not clear from the data. While several companies used multiple bases, most based their budget on other than sales, manufacturing costs, or direct labor hours. The expense budget and process capability were mentioned by two respondents as alternatives to the above.

PRODUCTION SYSTEM

The production system employed by the respondents resulted in either discrete products or mixed, i.e., continuous production and/or discrete products. The production level was established by most of the firms based on a sales forecast and for some a sales forecast in conjunction with market research and incoming orders. The resulting volume of production for 66% of the respondents involved either large batches or mass production of well established, stable, slowly changing products.

Two other related characteristics bear on quality control and its cost. They are labor cost and the number of production workers. The respondents indicated their labor costs as a percent of manufacturing costs ranged from 13% to 25%. Table 3 shows the ratio of production to quality control personnel employed by the respondents.

PRODUCT CHARACTERISTICS

Most of the respondents identified "appearance" type characteristics as the most important product characteristic affecting the quality control activity. Relatively few cited "functional" characteristics of the products as important. Concern for appearance has an impact on both manufacturing and quality control. The resulting effect is that the processes having the greatest impact on appearance characteristics become the most important from a quality control perspective.

Industrial orientation, operational climate, budgetary responsibility, and quality-sensitive product characteristics individually and collectively have

Table 3 Production to Quality Control
Personnel Ratio ($n = 10$)

Ratio	%
< 10:1	30
10−100:1	60
> 100:1	10

an influence on the allocation and expenditure of effort on product conformance
quality control. The nature of that influence is presented below as reflected
in selected expenditure data.

FINDINGS

These survey data depict manufacturer quality control behavior and priority
as reflected in expenditure practice. The quality cost data in Table 4 are organized
into the catagories of prevention, appraisal, and failure. Their total makes
up what has been defined as conformance quality control costs.

Theoretically, one would expect to see a distribution of expenditures on
prevention and appraisal activities skewed to over 50%, however, only 25%
reported spending over 50% of their total quality costs on prevention and
only 38% reported spending over 50% on appraisal. Failure costs exceeded
50% of total quality costs in nearly one-fourth of the companies that responded.
Based on these data — if they could be generalized — there is clearly much
to be done to shift the focus of attention toward defect prevention activity.
It should be noted that the previous study reported over one-half of the
manufacturers devoted in excess of 50% of their quality costs to the area
of failure and failure-related activities.[22]

The expenditure practice presented by industry classification gives additional
insight into the behavior of organizations with regard to product conformance
quality. The reported data are depicted in Table 5, which illustrates a substantial
variation in practice across the spectrum of industries represented.

Again a note of caution is appropriate. Even though the cost data collection
instrument was designed and tested to obtain compatible information from
all participants, semantics could have influenced the responses. Definitions
of prevention, appraisal, and failure may differ within a single company and
among people even when they are presumably fully informed. Such differences,
should they exist, could account for some of the variation reported here.

It is widely accepted in theory that quality activity should be oriented toward
defect prevention. Table 5 shows two noteworthy examples of this theory
in practice, namely paper (SIC 26) and fabricated metals (SIC 34).

Table 4 Expenditure on Prevention, Appraisal, and Failure Activity (Percent of Total Quality Cost and Percent of Respondents) $n = 8$

	Prevention		Appraisal		Failure	
	Cost (%)	Respondents (%)	Cost (%)	Respondents (%)	Cost (%)	Respondents (%)
	< 5	38	< 20	12	< 5	38
	5−9.9	25	20−50	50	5−50	38
	10−50	12	> 50	38	> 50	24
	>50	25				
		100		100		100

Table 5 Total Quality Cost Allocation by Industry
Industry Classification (SIC code)

Quality activity (percent of total quality cost)	26	28	34	35	36	38
Prevention	40	22	48	8	16	3
Appraisal	48	60	28	13	45	34
Failure	7	20	16	79	40	62

However, it should be noted that the two SIC 34 participants reported cost data at wide variance with each other. As noted above, there may be differences in what each classifies as prevention, appraisal, and failure costs. In this instance, the reported average data are not likely to be typical for the industry in general.

Chemical products (SIC 28) and electrical machinery (SIC 36) report the next highest percentage of prevention costs, although both are some distance behind the leading categories. The chemical-industry respondents reported the highest expenditures for appraisal activities; when those costs are added to those associated with prevention, a strong emphasis can be seen on reducing failures. Non-electrical machinery (SIC 35) and instrumentation (SIC 38) appear to provide a completely opposite picture. The largest portion of their costs are incurred in the failure area. Such a dichotomy might be attributed to the fact that each industry is represented by only one company, but one believes that each properly represents its industry class in general.

INDEXING

A final statistic of interest is the amount expended on quality in relationship to some widely employed index of corporate activity. Three such indices are manufacturing cost, gross sales, and value added.

Table 6 clearly shows that gross sales is the most popular base against which quality costs are measured. Manufacturing costs is the next most popular. Unadjusted value added is not as widely used as the former two. Total quality costs as a percentage of gross sales ranged from less than 1% to over 8%. Over one-half of the companies providing the data reported expending less than 5% of gross sales on conformance quality control. The previous study disclosed the same expenditure experience; 58% expended less than 5% on quality. The manufacturing cost and value added bases show quality costs representing a

Table 6 Total Quality Cost as a Percentage of Selected Bases

Industry (SIC code)	% of Manufacturing cost TQC	% of Gross sales TQC	% of Unadjusted value added TQC
20	*a*	.45	*
26	1	1	1
28	7	2	8
34	11	3.3	.5
35	7.4	5.3	18.6
36	*	6.5	*
38	12	8.2	25.6

*a*Data not available.
Note: May not add to 100% due to averaging or rounding.

much larger percentage of each. The significance of these various bases lies in the use that is made of the information. Obviously sales and marketing personnel find sales relationships most meaningful. Manufacturing management would find either manufacturing cost and/or value added measures more meaningful.

Table 7 depicts the components of total quality costs — prevention, appraisal, and failure—expressed as a percentage of the same bases used in Table 6. It follows that prevention and appraisal account for a substantially smaller portion of the measurement bases than failure costs.

Company conformance quality control expenditure practice for some respondents conforms to the theoretical framework underlying quality control management allocation decisions. The paper industry respondents offer an example of the theory in practice. Two responses from the fabricated metal industry — which when averaged together showed high prevention costs — showed a wide variation in reported prevention costs; it is unclear what the practice of similar business units of the same industry might be.

Although the chemical industry respondents reported a lower percentage of prevention costs, the total for prevention and appraisal was much higher than for failures.

Total quality costs as a percentage of gross sales is still the most popular measure. Companies probably should not rely on only one indicator. Multiple measures are highly recommended to meet the decision, making needs of the various members of management as they relate to product quality. Discussions with several of the respondents yielded other measures including quality cost

Table 7 Prevention, Appraisal and Failure Cost as a Percentage of Selected Bases[23]

Industry (SIC Code)	% of of Manufacturing cost			% of Gross sales			% of Unadjusted value added		
	P	A	F	P	A	F	P	A	F
20	a	a	a	a	a	a	a	a	a
26	.35	.25	.10	.25	.18	.07	1.35	.95	.40
28	1.7	4.2	1.2	.5	1.3	.4	*	*	*
34	.9	3.7	6.2	.3	1.1	1.9	.04	.17	.28
35	.6	1.0	5.9	.5	.7	4.2	1.6	2.4	14.6
36	*	*	*	1	3	2.5	*	*	*
38	.4	4.1	7.5	.3	2.8	5.1	.8	8.7	16.1

[a]Data not available.
Note: May not add to 100% due to averaging or rounding.

per unit of output. The specific measures employed reflected what the users felt would result in the most effective decisions.

One final observation is that the level of detail and precision with which quality costs are collected suggests the degree of significance attached to this information within the organization. A question that must be continuously asked and answered is whether the cost of quality cost data collection and analysis is justified by the benefits derived therefrom. To many it is of little importance. On the other hand, there is a sizable body of advocates making effective use of this data in the management of the organization as a whole, as well as the quality function in particular.

SUMMARY

This chapter has presented the economics of a major facet of product evaluation in a systems context. Quality cost allocation and expenditure reflects management policy and corporate behavior. Insight obtained from a study of conformance costs of related consumer product manufacturers was also presented.

RECOMMENDED READINGS

Blanchard, Benjamin S., Jr. *Design and Manage to Life Cycle Cost.* Portland, Ore.: M/A Press, 1978.

Brown, F. X. Quality Costs and Strategic Planning. *34th Annual Technical Conference Transctions.* Milwaukee: ASQC, 1980. pages 155-159.

Crosby, Philip B. The Management of Quality. *Research Management,* pages
    10-12.
Gilmore, Harold L. Quality of Employee Performance. *Quality Progress*
    (May, 1980): pages 14-17.
Pugh, Stuart. Quality Assurance and Design: The Problem of Cost Versus
    Quality. *Quality Assurance* (March, 1978): pages 3-6.
Quality Cost Committee. *Quality Costs — What and How.* 2nd ed. Milwaukee:
    ASQC, 1971.

NOTES

[1]Fremont E. Kast and James E. Rosenzweig, *Organization and Management,*
3rd ed. (New York: McGraw-Hill, 1979), page 18.

[2]Kenneth E. Boulding, General Systems Theory: The Skeleton of Science,
Management Science (April, 1956), pages 197-208.

[3]Robert A. Fetter, *Analysis for Production Management* (Homewood, Ill.:
Richard D. Irwin, 1957), page 9.

[4]Victor Lazzaro, ed., *Systems & Procedures: A Handbook for Business
and Industry,* 2nd ed., page 368.

[5]Roger G. Langevin, *Quality Control in the Service Industries* (New York:
AMACOM, 1977).

[6]Joseph M. Juran, Editor-in-Chief, *Quality Control Handbook,* 2nd ed.
(New York: McGraw-Hill, 1962), pages 1-2.

[7]*Ibid.*

[8]Adapted from Robert K. Fetter, *The Quality Control System* (Homewood,
Ill.: Richard D. Irwin, 1967), pages 2-5.

[9]William J. Masser, The Quality Manager and Quality Costs, *Industrial
Quality Control* (October, 1957), page 5.

[10]Juran, Handbook, pages 1-46.

[11]Warren R. Purcell, Quality Cost Control, *Industrial Quality Control*
(May, 1962), pages 22-26.

[12]Armand V. Feigenbaum, *Total Quality Control Engineering and Management* (New York: McGraw-Hill, 1961), pages 68-69.

[13]Timblin discusses the case where this curve is nonlinear. A Plan for Setting
Optimal Quality Performance Limits on Workmanship, *Annual Technical
Conference Transactions,* American Society for Quality Control (1966), pages
439-445.

[14]Leonard A. Seder indicates a nonlinear relationship is appropriate. How
to Evaluate a Company's Quality Control Need, *Annual Convention Transactions (1961),* American Society for Quality Control, pages 3-4.

[15]Charles Holt, et al., A Linear Decision Rule for Production and Employment
Scheduling, *Readings in Production and Operations Management,* Elwood
S. Buffa, Editor (New York: John Wiley & Sons, 1966), pages 442-478.

[16]It can be proven that the exact location of minimum total cost occurs where the absolute values of the slopes of the individual cost curves involved are equal for this is where a marginal increase in prevention and appraisal activity would be just equal to the resulting marginal decrease in failure activity.

[17]This statement carries the implied assumption that the quality assurance manager is minimizing costs subject to technical and management constraints related to the product, production system, quality level, business operating performance, and management policy.

[18]This theoretical concept is similar to allocation concepts proposed by other writers on the subject. Juran (Juran, Handbook, pages 1-10) depicts the occurrence of minimum total conformance quality cost control in the neighborhood of where cost of quality control (prevention and appraisal) equals the loss due to defectives (failure cost).

The cost model Timblin (Timblin, Performance Limits, page 441) developed depicts the same constituent cost relationships (prevention, appraisal, failure) as Seder discusses except that Timblin depicts a linear appraisal cost curve where Seder and the model in Figures 5 and 6 depict a nonlinear curve. Since Timblin's is a graphical model it depicts the cost curve relationships. A minimum total conformance quality cost, combined prevention and appraisal cost substantially exceed failure cost in Timblin's model.

Finally, Morgan also refers to the allocation of conformance quality control cost such that the combination of prevention and appraisal cost exceeds failure cost. Donald E. Morgan, *Some Models of Quality Cost Analysis for Industry, Technical Report No. 64-1*, Dept. of Industrial Engineering, Stanford University, Stanford, California (June 4, 1964), page 6.

[19]Philip Crosby, *Quality Is Free* (New York: McGraw-Hill, 1979), page 1.

[20]Harold L. Gilmore, Product Conformance Cost, *Quality Progress*, (June 1974), pages 16-19.

[21]Frank S. Leonard and W. Earl Sasser, The Incline of Quality, *Harvard Business Review* (Sept.-Oct., 1982), pages 163-171.

[22]Harold L. Gilmore, Product Conformance Cost, *Quality Progress*, (June 1974), pages 16-19.

[23]Tables 5 and 7 are not provided for comparative purposes. Cost data involved in each are from different respondents where there is more than one study participant.

Chapter 2

Corporate Policy and Human Resource Implications

Before any PT&E accomplishments can be realized, firm policies and procedures relating directly to the PT&E function have to be formulated. In the present chapter we discuss the formulation of policy and procedures that, when implemented, impose a responsibility on the PT&E management to become intimately familiar with company objectives. Conversely, it also requires corporate management to be familiar with PT&E capabilities. A most effective way to achieve this management rapport is through formal policy and procedures, which provide a written document of the scope of responsibility and the degree of authority. A definition of the PT&E responsibilities and capabilities is a prerequisite to our discussion of policy statements, position descriptions, and personnel requirements, which are the basis for a successful and meaningful operation.

PT&E RESPONSIBILITIES AND CAPABILITIES

An essential test and evaluation management contribution is to provide senior management with data on the PT&E function. This should be in the form of general test and evaluation philosophy and what is expected of it in terms of current business and future requirements and what specific services it can perform.

Since test and evaluation is in a rapidly changing growth period, management must be informed of the current status and its implications. Failure to do this would detract from the overall organizational approach to test and evaluation and, therefore, from its ultimate effectiveness. This is essentially a communications problem between middle and top management. The latest test and evaluation requirements and responsibilities must be presented, either orally or in writing, to keep top management informed.

The very existence of a test and evaluation organization depends on the service that it will provide. It therefore becomes essential to indicate clearly what specific contributions will be made, in what areas, and how they will be accomplished. For instance, if the PT&E function is to establish the product warranty period, it should present its approach and methods to be used in as simple and straightforward a manner as possible.

POLICY STATEMENTS

What is policy? Most inquiries made during the preparation of this book, relative to policy statements, were not fulfilled. Several reasons were given, the most frequent one being that none existed. Another important reason, revealed by an American Management Association survey[1] in 1962, was that 110 participants out of several hundred found it difficult to define the word "policy." Additional evaluation of the survey results disclosed that many "policies" were actually procedures, practices, or rules, thus indicating the differences that exist in understanding their meaning and importance.

There are many definitions of policy, but the only brief, unperplexing one is: A policy is a guide for action.

Policy statements establish a specific course of action to govern the operations of an organization, expressing its philosophy, principles, purpose, and methods of attaining the test and evaluation goals.

Among existing definitions, the following three deserve particular attention.

A policy is a verbal, written, or implied overall guide setting up boundaries that supply the general limits and direction in which managerial action will take place.[2]

A policy is basically a statement, either expressed or implied, of those principles and rules that are set up by executive leadership as guides and constraints for the organization's thought and action. Its principal purpose is to enable executive leadership to relate properly the organization's work to its objectives.[3]

A policy is a definition of common purposes for organizations, components, or the company as a whole in matters where, in the interest of achieving both complete and overall company objectives, it is desirable that those responsible for implementations exercise discretion and good judgment in appraising and deciding among alternative courses of action.[4]

It is essential to understand that policies provide a means for carrying out the management process and, therefore, assist in the decision-making process and can only be formulated by top management.

One of the most important contributions that test and evaluation management can perform, then, is to clearly define and indicate to senior management the particular role that the test and evaluation function will play in achieving the corporate objectives and contractual requirements. Occasionally a test evaluation activity is "born" out of necessity to fulfill a contractual obligation. As a result, the ensuing effort amounts to nothing more than an academic exercise with little impact on the end product. What is needed is a policy statement accompanied by management job descriptions. This can be accomplished by formulating and recommending test evaluation methods and procedures for management endorsement. Once the policy and position descriptions are officially approved and established, test and evaluation functions and responsibilities can be effectively discharged.

The importance of test and evaluation policy arises out of at least three basic relationships between it and the product development program:

1. Since policy states the intentions of the program, its clear definition and broad, uniform understanding are essential to consistent action. This is especially true in a large corporation in widely dispersed operations or in a multidivision/multiproduct firm. Such an environment will probably lead to the development of separate policies with resulting confusion, inefficiency, and variations in performance if not carefully guarded against.
2. Sound policies are an essential basis for sound practice. The resultant output will inevitably reflect shortcomings inherent in the policy.
3. Policy is the essential yardstick by which to measure accomplishment in the program. A program cannot be measured for accomplishment without understanding what it was designed to accomplish. Hence, appraisal of the entire test and evaluation program — in product development, performance evaluation, consistent quality level, design improvement, satisfactory service life, and throughout all test and evaluation functions — must rely largely on explicit declarations of policy as measuring sticks. Such appraisal consists essentially of comparing practice with policy and the results of practice with both policy and practice. In the absence of policy, such comparisons cannot be made.

POLICY PRINCIPLES

Several principles, with respect to statements of test and evaluation policy, are described below.

Explicit Policy. Test and evaluation should be clearly stated in quantitative terms so that there is no question as to what it involves.

Policy Dissemination. Policies must be communicated to all who are responsible for seeing that they are implemented. To this end, specific means must be provided for informing all levels of appropriate policy provisions.

Uniformity of Content. Complete uniformity is probably impossible in a large organization, but basic policy should be as uniform as possible throughout. Variation in policy will lead to confusion and interpretation.

Conformity of Interpretation. Interpretations should be defined in detail and then given wide dissemination so that all may be informed.

Appropriateness. Policies should be appropriate to the situation. They must be prepared specifically for the activity involved and all inclinations to copy policy from another organization must be conscientiously avoided. Only the most fundamental and general policies can be seriously considered for adoption. Differences in company size and in program and product orientation will require that policies be "tailor made."

Reasonableness. Specific test and evaluation policies must be reasonable in terms of overall corporate objectives; they must be consistent with customer requirements and consistent with governmental and legal requirements. In the final analysis, the policies must contribute to the ultimate success of the product in the marketplace. A policy that does not satisfy these requirements is inappropriate.

Scope. The policy content may be as general or as detailed as you wish. No all-inclusive solution can be offered, since it is largely up to the individuals and companies involved. Generally the larger companies develop the most extensive and detailed statements. The next section outlines a representative policy statement that reflects the foregoing principles.

A TYPICAL PRODUCT TEST AND EVALUATION POLICY

Introduction

This policy statement establishes the PT&E and related reliability, safety, maintainability, and human engineering policies that are mandatory within the company.

In implementing these design assurance concepts, attainment of the principal factors of reliability, safety, maintainability, and human engineering will provide assurance of optimum product effectiveness in conformance to customer requirements.

The basic PT&E policies set forth are established and maintained by the Product Test and Evaluation Department. It is the responsibility of the department to keep the policies up to date and to rule on all requested changes.

Objectives

It is the policy of the company to attain optimum product effectiveness in the development, manufacturing and end use of its products. The objectives are summarized as follows:

Reliability, maintainability, safety, and human engineering shall be major considerations in all programs and operating decisions and shall be established for application to all applicable products to ensure that prescribed and/or basic levels of effectiveness are met. These will include procedures for detecting and correcting causes of unreliability, poor maintainability, unsafe conditions, and unsatisfactory man/machine interfaces.

Suppliers and subcontractors shall be required to participate in PT&E programs.

A continuing program in PT&E training shall be conducted among all levels of personnel to assure the attainment of these objectives.

Definitions

Product test and evaluation (PT&E) is a management philosophy that embraces all company product design characteristics with reliability, safety, maintainability, and human engineering being principal factors.

Reliability is a measure of the time stability of a product's performance. Technically, it is expressed as the probability that an item will perform its intended function satisfactorily for a specified length of time when used for the purpose intended and under the condition for which it was designed to operate.

Maintainability is a characteristic of design and installation which is expressed as the probability that an item will be retained or restored to a specified condition within a given period of time, when the maintenance is performed in accordance with prescribed procedures and resources.

Safety applies engineering data to the design to ensure the conservation of human life and its effectiveness and the prevention of damage to items or associated equipment.

Human engineering is the area of human factors that applies scientific knowledge to the design of items to achieve effective man-machine integration and utilization.

Responsibility for company reliability, safety, maintainability, and human engineering is delegated to the Product Test and Evaluation Department. This department has been established to solidify company effectiveness capabilities and to effect coordination with engineering, manufacturing, and product assurance operations as a means for reaching mutual understanding and assistance on points of common interest. At the same time, a reservoir of unique effectiveness capabilities and experience is available.

The Product Test and Evaluation Department is responsible for the following specific activities:

1. To integrate company effectiveness programs.
2. To assist in preparation of technical proposals and cost estimates for future programs, as dictated by customer needs and/or interests.
3. To assist marketing in approaching customers on new business and in the determination of specific effectiveness needs.
4. To keep abreast of military and commercial developments in reliability and product effectiveness and to keep the company in a prominent position in the field through active participation in government and industry activities.
5. To maintain and/or improve the operational effectiveness of the company PT&E functions through programs of training, cost control, personnel recruitment, and performance specifications.
6. To review and approve annual plans and budgets of the product test and evaluation functions prior to their submission to the General Manager.
7. To approve formal policy statements of other company and project areas that directly affect the technical activity of the product test and evaluation function.

PRODUCT TEST AND EVALUATION FUNCTIONS

Reliability

A basic Reliability Program shall be incorporated into all PT&E programs to the extent necessary to assure compliance with customer and/or legal requirements. This basic Reliability Program shall consist of, but not be limited to, the following major considerations:

1. Analysis, interpretation, and preparation of product effectiveness specifications.
2. Advance planning and research.
3. General service and support to other departments.
4. Design review and approval.
5. Establish reliability goals.
6. Determine preliminary prediction of reliability.
7. Implementation of training and indoctrination programs.
8. Supplier selection, surveillance, and control.
9. Program surveillance and monitoring.
10. Test planning and evaluation,
11. Reliability mathematics and statistics.
12. Reliability demonstration testing.
13. Component evaluation and application studies.
14. Data exchange program.

15. Data processing and reporting.
16. Failure mode and effects analysis.
17. Field data feedback and corrective action.
18. Approval of change requests and change orders.
19. Packaging and shipping procedures.
20. Establish reliability design criteria.

Maintainability

A basic Maintainability Program shall be incorporated into all PT&E programs to the extent necessary to assure compliance with customer and/or legal requirements. This basic Maintainability Program shall consist of, but not be limited to, the following major points:

1. Analysis, interpretation, and preparation of proposal and contractual specifications for maintainability.
2. Advance planning and research.
3. General service and support to other departments.
4. Design review and approval.
5. Establish maintainability goals.
6. Determine preliminary prediction of maintainability.
7. Implementation of training and indoctrination programs.
8. Supplier selection, surveillance, and control.
9. Program surveillance and monitoring.
10. Test planning and evaluation.
11. Maintainability mathematics and statistics.
12. Maintainability demonstration testing.
13. Field data feedback and corrective action.
14. Approval of change requests and change orders.
15. Packaging and shipping procedures.
16. Establish maintainability design criteria.

Safety

A basic Safety Program shall be incorporated into all PT&E programs to the extent necessary to assure compliance with customer and/or legal requirements. This basic Safety Program shall consist of, but not be limited to, the following major points:

1. Analysis, interpretation, and preparation of proposal and contractual specifications for human engineering.
2. General service and support to other departments.
3. Design review and approval.
4. Establish human engineering design criteria.

5. Perform human engineering analysis.
6. Perform human error analysis.
7. Program surveillance and monitoring.
8. Approval of change requests and change orders.
9. Test planning and evaluation.
10. Implementation of training and indoctrination programs.

Requirements

The requirements of the Product Test and Evaluation Program are mandatory for all personnel, processes, and equipment regardless of whether the functions are performed within the division, corporation or at a vendor facility. When customer required, product effectiveness data shall be made available to bona fide representatives.

DESIGNING POSITION DESCRIPTIONS FOR THE PT&E OPERATION

The second facet of the management definition is position descriptions. Here, as with the policy statements, there is a wide variation in what is considered acceptable and necessary. Many corporate executives maintain that position descriptions are not practicable for the simple reason that any position reflects the interests, strengths, and weaknesses of the person performing it. In many instances this is true, but there are also those who believe that some degree of standardization is necessary and that the position description is an effective vehicle for accomplishing it.

Another benefit derived from the preparation of position descriptions is that the process of writing the descriptions proves to be an unplanned-for dividend. The requirement to write an accurate description forces thinking that, in turn, uncovers organizational flaws, open communication loops, and erroneous job emphasis.

Management concepts (test and evaluation concepts included) impinge on the position descriptions of the functional supervision. A review and evaluation of position descriptions can provide deep insight into top management's approach to a given area. The following position description illustrates how top management concepts are implemented and reflected in individual positions.

It is well known and appreciated that the most efficient approach to producing a quality product is to have accurate and complete specifications. The same notion is valid for individual performance. The obvious way to establish clear performance requirements and limitations is to prepare accurate and detailed job or position descriptions. In this way the requirements for each job are carefully identified and will encourage improvement and study by aspiring employees.

The requirements that should be considered in developing descriptions include physical traits, personality, skills and knowledge, responsibilities, authorities, and reporting relationships. The former requirements pertain more to jobs at the technical level with responsibilities, authorities, and reporting levels applicable to supervisory and managerial positions.

The following position description demonstrates the scope and contents of positions which may represent those responsible for the test and evaluation function in a variety of industries.

PRODUCT TEST AND EVALUATION MANAGER

Duties

1. Develops for approval, PT&E organization policies; work systems and procedures; and budget forecasts of manpower, facilities, and expense.
2. Confers with management, customer, subcontractor, and vendor representatives in planning, scheduling, coordinating, and directing control programs in conformance with customer requirements, engineering specifications, and production schedules.
3. Directs, through key supervisory personnel, employees engaged in:
   a. Engineering, designing, and developing special test tooling programs.
   b. Generating estimates of the quality control portion of quotation estimates.
   c. Reviewing and approving engineering test specifications.
   d. Studying, evaluating, and developing test methods, techniques, and procedures.
   e. Planning and scheduling the manufacture and procurement of test equipment.
   f. Calibrating and maintaining test equipment; operating and maintaining measurement standards for the control of all instruments used in quality control and manufacturing.
   g. Performing prototype tests of new products to eliminate production problems, functional and environmental tests, special sampling tests, in conformance with product drawings and specifications, military specifications, and standard practice instructions.
   h. Inspecting parts, materials, tools, equipment, and products; reviewing and approving inspection fixtures and gages designed by manufacturing engineering.
   i. Developing, establishing, maintaining, and controlling an effective PT&E program to facilitate detection, correction, and prevention of defects in materials, parts, assemblies, and products; evaluating and monitoring vendor and subcontractor quality performance.
   j. Establishing and maintaining programs and procedures to insure that devices perform their intended functions within specified time limits

under environmental conditions. Specific responsibilities include providing mathematical models, component reliability indices, and reliability standard practice instructions for design and acceptance procedures; scheduling design review audits and analyses; establishing vendor reliability ratings, devising environmental studies for parts and components; and performing probability analyses of failure.

k. Preparing, maintaining, and distributing quality control and safety Standard Practice Instructions.

l. Evaluation of all test data for conformance to requirements.

m. Reviewing all capital requests for test equipment and maintaining accountability for customer property and capital equipment.

4. Trains, motivates, develops and counsels personnel under his or her direction to improve job performance and insure effective, economical operations.

5. Maintains harmonious employee relations and assists subordinates in the handling of grievances; participates in customer negotiations and changes as they affect activities under his or her direction.

6. Appraises employee performance and gives intermediate approval on recommendations for personnel procurements, salary increases, promotions, demotions, and dismissals in accordance with work requirements and within authorized limits.

PLANNING FOR PT&E HUMAN RESOURCE REQUIREMENTS

Human resource considerations generally revolve around technical requirements and the quantity of people available. Technical requirements are largely a function of the product to be tested, the facilities to be operated, and the authority and responsibility to be delegated. The constant advances in research and development technology and product complexity have resulted in increasingly complex techniques and methods in test and evaluation. As a result, constant pressure is exerted for increased technical capability in test and evaluation personnel. No longer will the gadgeteer suffice in the laboratory. The techniques of testing require scientists and engineers. Industrial management is beginning to recognize the need for professional services, and many corporations that produce sophisticated high reliability products have already recognized the need. Their approach will be followed in time by smaller concerns, which will staff themselves accordingly or seek professional test and evaluation services from independent laboratories. We shall now discuss the personnel requirements of a test and evaluation function.

Personnel Requirements

The test and evaluation organization should consist of personnel who are qualified to meet the job-description requirements. An extensive test and evaluation organization within a system manufacturer would include personnel who are thoroughly experienced in the following background areas:

Systems engineering
Component engineering
Specification review
Reliability/maintainability/human factors/system safety engineering
Design and analysis of statistical experiments
Test facility design and operation
Test planning and environmental testing
Data collection, evaluation, analysis, and follow-up
Failure analysis—physics of failure
Manufacturing processes
Market research

From this list, the variety of background experience, in the depth required, exceeds the capability of a single individual. Product test and evaluation is a team effort, and the task of management is one of dealing with many sorts of individuals and of insuring high competence in all of these background areas. And, above all, management must see to it that these many specialized practitioners perform as a team dedicated to a common objective. Let us discuss each of the background areas in sequence.

Systems Engineering. Capability to perform an evaluation of the product as a whole requires a person who is able to design system-level tests and to design subsystem and component tests that accurately reflect the system application. The individual must have an appreciation for the technical complexities introduced by the equipment interfaces, operating conditions, and final use environment. To achieve optimum economy of effort, maximum use of subsystem and component data for system performance evaluation purposes must be achieved. This requires a great deal of technical sophistication.

Component Engineering. Just as a chain is no stronger than its weakest link, a system is no more reliable than its most unreliable component. Therefore, much of the test and evaluation work is performed at the component level. Extensive background in component performance capability and intended application is necessary in order to design meaningful tests and to perform valid data evaluation.

Specification Review. Specifications define the product in terms of performance capability. An integral part of the test and evaluation function is to evaluate the product relative to the specification requirements. Where a difference exists, specific action is required to change either the specification and/or the product design. In addition, test and evaluation personnel should provide environmental and test information for incorporation in specifications.

Reliability/Maintainability/Human Factor/System Safety Engineering. The accent today (and for the foreseeable future) is on product effectiveness. Test and evaluation personnel should be keenly aware of the product requirements in this area and of the technical requirements of the disciplines that they represent. Specific test objectives may be solely in this area and, in most all cases, data can be derived from all testing when properly designed for product effectiveness purposes. The requirements to be considered are quite different. They go far beyond the point of determining whether something works; they consider frequency of failure, margins of safety, ease of maintenance, and compatibility with humans.

Design and Analysis of Statistical Experiments. The value to be derived from statistical experimentation is well known to the quality control practitioner. However, this value is not so well known to the test engineer and general management. Not only can very significant cost and time savings be realized through the application of existing techniques but the objectives of product effectiveness can also be served.

Test Facility Design and Operation. Naturally, the capability must exist to operate the various test and environmental facilities and to design new facilities and fixtures to meet ever-changing test requirements.

Test Planning and Environmental Testing. Skill must exist to permit the preparation of adequate test plans and procedures. Knowledge of test objectives, mission application, method of data evaluation, and the ability to write step-by-step procedures leading to the accumulation of valid test data are essential.

Data Collection, Evaluation, Analysis, and Follow-Up. The actual testing (which is only part of the product evaluation function) in order to enable proper decisions to be made and the follow-up to those decisions represent the more significant aspects of product evaluation. Knowing what data to collect and in what form to collect it to facilitate its evaluation is of prime concern. A subsequent chapter provides further insight into the data-collection problem.

Failure Analysis—Physics of Failure. Failures during the course of testing are a certainty. Without a detailed analysis of the mode and mechanism of each failure experienced, design maturity or product improvement will not be realized. Sophisticated analysis techniques will be required in order to properly evaluate the physics of failure. The results of these analyses also need correlation with associated reliability analyses.

Manufacturing Processes. Many of the problems encountered during the test program will reflect on the manufacturing processes and procedures. To assess the problems properly and to propose meaningful corrective actions, a familiarization with the processes employed is necessary. In addition much testing will be performed specifically to assess the effects of process variations on product performance.

Market Research. Products must meet the tests of the marketplace. Individuals skilled in the design of valid tests to determine market acceptance or rejection of products is essential. Results from these tests will be the basis for further product development or cancellation.

Stated again for emphasis, the technical requirements for product test and evaluation are no less demanding than the engineering talents required for the end product design task. When we consider the design implications of applying the general rule—that is, that the inspection device must be ten times more accurate than the item inspected—to the job of the test engineer, the requirements placed upon him become quite demanding.

SUMMARY

Our human resources are our most valuable asset. However, under the more obvious demands of facilities, schedules, and finances, human resources frequently are not given adequate attention. PT&E is not a program of instruction in universities or technical schools, thereby requiring on-the-job development. Perhaps the growth of PT&E will result in more formal attention being devoted to this engineering task. Until such a time, industrial management must provide for its requirements by developing on-the-job skills. For the benefit of the persons charged with this responsibility, this chapter has provided an outline of the general characteristic requirements that should be met by practitioners in the field. It also has stressed the importance of a formal approach to policy definition. The position descriptions, when applied, will be of significant aid in the determination of personnel requirements, and they will help in the ultimate selection of training of personnel.

RECOMMENDED READINGS

Bennet, C. L. *Defining the Manager's Job.* American Management Assoc. Research Study No. 33. New York: American Management Assoc., 1958.

Evans, Gordon H. *Managerial Job Descriptions in Manufacturing.* American Management Assoc. Research Study No. 65. New York: American Management Assoc., 1964.

Miller, Ernest C., ed. *Human Resources Management: The Past is Prologue.* New York: AMACOM, 1979.

Beach, Dale S. *Personnel: The Management of People at Work.* 3rd ed. New York: Macmillan, 1975.

NOTES

[1]Valliant M. Higginson, Management Policies I, Their Development as Corporate Guides, AMA Research Study 76.

[2]George R. Terry, *Principles of Management*, 4th ed. (Homewood, Ill.: Richard D. Irwin, 1964), page 278.

[3]Ralph Currier Davis, *The Fundamentals of Top Management* (New York: Harper and Brothers, 1951), page 13.

[4]*Professional Management in General Electric*, Book Two, *General Electric's Organization* (General Electric Company), page 15.

Chapter 3

Organizational Methods and Financial Implications

Chapter 2 dealt with the essentials of policies, procedures, and human resource considerations necessary to perform the PT&E function. This chapter discusses the position of the PT&E department in the organization and the economics involved in test program implementation. The most important resource of any business is its people. As Peter Drucker said, "Business is a human organization, made or broken by the quality of its people."[1] Therefore, the effective placing of people is of primary importance to business success and to the test and evaluation function in particular.

PT&E FUNCTIONS

Two clearly defined basic functions are almost always found in business firms: production and marketing. Another function, not so clearly defined, is the determination of product performance characteristics and quality. Unfortunately, this determination is fragmented into various specialties such as quality, reliability, safety, inspection, marketing, and human engineering. Today the need exists for a system approach to PT&E. Considered in this context, the PT&E function is to evaluate the product for market acceptability, production readiness, and customer use. In this light, PT&E warrants a position in the decision process equal to and linking production and marketing. Figure 1 indicates the schematic flow of activities involved in test and evaluation in its most concise form.

The provision of products and services is the function of production. Market research, selling services, and the distribution of products are the functions

Figure 1 PT&E activities flow chart.

of marketing. The functions of PT&E management are to plan, organize, direct, and control the necessary activities to provide test and evaluation services which augment the activities of the other two.

Fragmentation of the responsibilities mentioned above evolved from the long-run trend in job specialization. The larger the organization, the greater the degree of specialization and functional fragmentation and the greater the tendency to lose track of the overall organizational objectives. A PT&E systems management approach would encourage just the opposite effect. It would facilitate a balancing of interdepartmental advantage and disadvantage to seek an optimum system. Benefits to be derived would include reduced cost from a variety of sources, more efficient management controls, a greater appreciation for product performance throughout the firm, and a fuller participation in the marketing and operating decisions of the firm by PT&E personnel.

In any PT&E activity, the first concern of the PT&E manager is to provide inputs: test specimens, test and inspection equipment, operating supplies, and human resources. Once the inputs are provided, product effectiveness can be evaluated relative to performance criteria. This is the stage to which the test and evaluation manager gives the most attention. Scheduling tests on machines, assigning men to jobs, controlling test quality, improving test methods, developing new techniques, and the feedback of results within the firm are among the activities that the manager must focus on during evaluation. The final stage of PT&E is the completion of outputs or the completion of the evaluation. These results are provided to production, marketing, and engineering to aid in the economic design, production, and distribution of the product.

THE PT&E SYSTEM CONCEPT

A PT&E system is a framework of activities within which product evaluation can occur. At one end of the system are inputs and at the other end are outputs.

Connecting the inputs and outputs are a series of evaluation operations in the form of a flow process. Figure 2 represents a simplified PT&E system.

The evaluation of any product or service can be viewed in terms of a PT&E system. For example, the manufacture of television sets uses such inputs or materials as cabinets, picture tubes, printed circuits, tuners, and a variety of small components. After the initial design concept is released, the inputs must be evaluated before production at various stages of assembly right through the finished product. After each item and each stage of production are evaluated and appropriate changes made, the design is released to the Production Department. Based on the experience gained during the preproduction phase, the PT&E production phase is implemented. Final inspection is performed prior to shipment or finished goods storage, resulting in either product acceptance or rejection. Once the product is in the user's hands, there is no assurance of further performance feedback. However, customer complaints must be answered and the resultant product performance data must be evaluated and acted upon. The production of boats, automobiles, airplanes, and missiles encompasses test and evaluation. This same concept applies to service organizations, such as hospitals, railroads, and supermarkets. All involve inputs, all pass through various stages of evaluation before going on line, all must pass final inspection, and all must pass the acid test of ultimate user acceptance.

Product test and evaluation, when viewed from a system concept, has universal application and logical foundation from an organizational standpoint.

PT&E CONTRIBUTION TO THE COMPANY

Structured as envisioned above PT&E benefits accruing to the company are significant. From an economic point of view, test and evaluation helps to prevent overdesigning the product but makes the product as reliable and safe as it needs to be. In addition, it provides the basis for determining the warranty and forecasts the product's cost. As a marketing tool, test and evaluation results prove company competence in the pursuit of new business and facilitates product differentiation on the basis of quality, reliability, safety, and maintainability. Test results support new product-development decisions as a planning and economic aid and are of immeasurable value in proving that products meet performance requirements in the event of lawsuits.

ORGANIZATION OF PT&E

Organization is largely a matter of isolating the specialties required by the organization and providing them. This is embodied in the definition of "organize," which means to arrange or constitute in interdependent parts, each

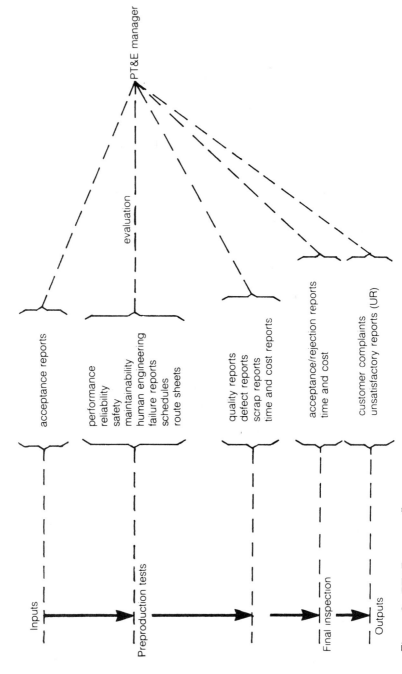

Figure 2 PT&E systems flow.

having a special function in relation to the whole. The need for organization units, each doing its own specialized work, arises because work must be performed by competent individuals and in manageable units capable of being handled by an individual. Modern organization is often called functionalization and is also frequently used synonymously for department, division, or group, implying a relationship to the whole. The key element in the approach to this division of work is the job evaluation and job description. The functional requirements must be evaluated in the light of the "average" practitioner and the job described to provide for such an individual.

By proceeding along these lines it is possible to divide the work of a complex operation into manageable units and avoid the unwise situation of delineating managerial functions or specializations of greater magnitude than most men can handle. One cannot rely on finding the "exceptional" man.

An outgrowth of functionalization is the need for coordination. This is dealt with by means of the relationships that each department head is required to maintain. These relationships are imbedded in job descriptions and are achieved through the personal leadership and policy statements of management and informal working arrangements. The specific relationships designed into the organization must be responsive to the dominant needs of the operation. This means that where timely decisions are a requisite for successful operations, direct relationships from that area to the chief policy making body must be provided. Coordination is also enhanced by the chief executive via a selective upward flow of operating results and the resolution of conflicts between functional areas.

LINE AND STAFF

The clarification of relationships makes a distinction between line and staff units necessary. However, when it comes to deciding which components of an organization are line and which are staff, the distinction is not always easily drawn. Basically, all of the specialized units that carry out the main operating activity of the organization are line, and all of the units that act in a service or advisory relation to the line are staff. Perhaps a clearer illustration (although not realistic) can be obtained by evaluation, on a unit-by-unit basis, of whether "business" could be carried on without the unit in question. If the answer is yes, then the activity is staff to the operation. The line may also be distinguished by another basic characteristic. This is the right to hire and fire. The line superior has the right to impose discipline or to discharge. The staff has this right only with respect to subordinates within their own functional specializations—that is, persons with whom they have a line relationship.

The key consideration requiring line-staff differentiation is authority. The staff cannot be placed in a position or allowed to give orders or take action that will effect the line units. This simply cannot be allowed. All direct orders must come from one's line superior and from no one else.

Two other points about line and staff require mention.

1. Since the chief executive is head of both line and staff, one of his principle duties is to coordinate them and to settle disputes. This responsibility was mentioned earlier. He should not be thought of as being line-oriented only. He must be thought of as being the primary coordinator of line and staff.
2. As alluded to earlier, an individual may be line in his relation to one person and staff in his relation to another. PT&E provides an illustration of this situation. Production regards it as staff, although the inspection supervisor has a line relationship to the PT&E manager. The nature of the line and staff roles clearly indicates the need for their accurate definition to those individuals involved; otherwise, confusion with respect to responsibility, authority, and rank will detract from operational effectiveness.

DECENTRALIZATION

Decentralization is another characteristic of business organizations. The extent to which it is employed varies widely in practice. Delegation determines the degree of decentralization of authority, and the degree of decentralization does *not* determine the organizational form of the organization. A one-facility operation can employ as much decentralization as a multi-facility operation. The main objective is to gain organizational initiative and responsiveness by increasing the number of rapid, but well-informed, decisions made outside of headquarters. It requires not only delegation of authority but competent subordinate managers and effective performance data at the level of decentralization. From an organizational point of view, decentralization requires many more capable managers than a highly decentralized organization. Unless decentralization is introduced carefully, coordination and efficiency will suffer a serious loss. A highly centralized operation, on the contrary, is likely to depend excessively on one or two key people who manage everything. The advantages and disadvantages of each must be carefully weighed in light of the specific work environment before adopting a particular approach.

DIVISIONALIZATION

Divisionalization of the line is the counterpart of functionalization in the staff. The staff is divided into personnel, finance, and others because the chief executive cannot perform all of these functions. Divisionalization of the line activity occurs for the same reason: the chief executive cannot supervise each individual line unit for many reasons. It is an effort to reduce line operations of a company to manageable units, each in charge of a separate executive to eliminate multiple supervision from the chief executive. The failure to divisionalize a growing operation at the point at which division is required to provide effective control and direction is a prelude to operating troubles.

Divisionalization may take a variety of forms, (for example, product, project, and geographical), which may be found in combination and may be carried to the ultimate extreme by establishing independent profit centers. An important implication of divisionalization, in whatever form, is that it usually involves decentralization. Therefore, management must be prepared to provide for the associated requirements related to decentralization mentioned above.

APPLICATION TO PRODUCT TEST & EVALUATION FUNCTION

The concept of PT&E, as we have presented it, creates a need for coordination and direction. In the broadest sense this need is usually dealt with by means of management relationships, which are defined in terms of who reports to whom.

These relationships vary according to the nature of the business—for instance, businesses dealing with the government find it advantageous (if not absolutely necessary) not to relegate the PT&E function to a situation in which it "reports" to anyone other than "top management." On the other hand, businesses not influenced quite as much by any one specific customer may place PT&E under the manufacturing or operations executive. The latter situation occurs whenever the customer does not have such a direct bearing on the success or failure of the business. It is particularly important that the location be initially selected to suit the dominant needs of the operation. By observation, then, one can almost gage the impact and importance of PT&E by its location on the organizational chart. However, a word of caution: although the organizational chart provides the formal relationship, frequently it is the informal relationships that have the most influence on day-to-day operations; therefore, it is extremely important to bear this aspect in mind and to ascertain that the two are in harmony with each other. Failure to take this precaution may lead to gross misconceptions in the evaluation of the test and evaluation influence. The differentiation of line and staff was discussed above. When organized as staff, the PT&E function will be confined to providing service, advice, and product performance information. As a line function it will have responsibility and authority to contribute to the outgoing product flow by permitting shipment, holding up production, or causing changes in product design.

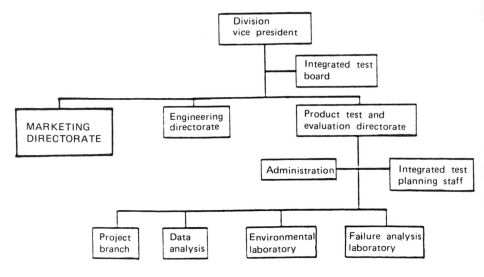

Figure 3

We have explored the organizational characteristics of modern, large businesses. We have seen that the business is nearly always a corporation, that it is functionalized, and that it has line and staff distinctions. Now let us assess some of these concepts as they exist today by reviewing a typical organizational chart.

The organization of a typical company (Figure 3) provides an example of a centralized test facility established to provide overall division test support. In addition, it furnishes the method for integrating and coordinating the total test program on any specific project. It is interesting that this group has the authority to endorse or reject the design through its position on the Integrated Test Board that is established for this purpose.

Testing is handled on a project basis, with facility requirements coordinated at the directorate level by the administrative staff. The elevation of the test board to the divisional staff level requires equal representation of marketing, test, and engineering personnel. If agreement cannot be reached by the board, resolution must come from the division management.

OPTIMUM LOCATION

Where should PT&E be placed in the organization? Should it be line or staff? What degree of specialization is needed? These valid questions must be answered wherever the test and evaluation is conducted on a formal, organized basis. We do not answer those questions because no outsider can dictate where the function should be specifically placed, or whether it should be line or staff. These decisions will depend on specific situations that exist in the firm.

Generally, the PT&E function should be placed where it can most effectively achieve its objectives consistent with the overall objectives of the firm. This will minimize the likelihood of suboptimization problems. As to line and staff, this will depend on whether or not it is authorized to give direction. If it can shut down the production line or prevent shipment, it will be a line function. If it can provide advice so that someone else can make such a decision, it should be staff.

There is still another alternative: combining line and staff. This involves dividing the function, but it could be considered as a possibility.

FINANCE

The financial implications of product test and evaluation are significant. For example, equipment designed to simulate the environmental extremes of space travel can cost a million dollars or more. A missile checkout facility will cost millions of dollars. The development and use of costly facilities like these are not restricted to aerospace research and development. For instance, an automotive test facility claimed to be an example of "fallout" from aerospace R&D into the automotive field, costs around $500,000 today.

Capital investments of this size and expense always require top management approval. Taxes, depreciation, interest on investment, and general and administrative expenses are significant, and directly affect corporate earnings. Justification for such facilities must reflect consideration for the overall long-term business objectives of the firm, the financial burden imposed, and the expected degree of utilization. Test equipment utilization has a direct bearing on unit costs in the same way and for the same reasons as production equipment utilization. The ideal situation, of course, would be to attain 100% equipment loading, although this cannot be attained in practice except for limited periods of time.

SCIENTIFIC MANAGEMENT TECHNIQUES

To a certain extent, some of these management techniques have been applied to test and evaluation operations. Because of the costs represented by the test function, more comprehensive application of the techniques is warranted. An excellent treatment of the basic ideas and general concepts for planning and scheduling industrial products is provided in a book by Elwood S. Buffa.[2] The application of these concepts to test programs is well worthwhile. As an example, consider the acquisition of additional environmental test facilities. A serious mistake would be to proceed with procurement without an economic study, only to find out too late that an error had been made resulting in an undesirable capital risk.

CAPITAL BUDGETING, PLANNING, AND CONTROL

The investment associated with test and evaluation as stated earlier can be (and frequently is) substantial. Therefore, a clear understanding of the general concepts associated with capital budgeting control is necessary. Whatever the method ultimately adopted, it will be based on the balanced application of the following four characteristics:

1. A clear perception of the future financial needs of the test and evaluation function.
2. A disciplined approach to both short- and long-range financial planning.
3. An organized approach to financial planning and control, tailored to the task to be accomplished.
4. A closed-loop system for the collection, analysis, and evaluation of relevant financial information.

The key concept is to make capital budgeting decisions (for example, planning expenditures whose returns extend beyond one year) more effective by emphasizing the development of systematic procedures and rules for developing, evaluating, selecting, and implementing various capital investments.

Test and evaluation management must take a disciplined approach to financial planning. Planning, whether for the short or the long range, is not a one-shot effort. Management must establish financial planning as a continuing, integral aspect of the management function. Through the planning activity (long range), objectives may be defined in such a way as to permit their objective evaluation with respect to all appropriate alternatives. The forward planning time span should reflect the longest period covered by any one of the projects being considered. (Generally, financial planning should project into the future as far as it is reasonable to do so and, at the same time, produce accurate, useful guidelines for action.) In essence, the planning task is to identify the results desired and to establish the means for attaining them. In this way, management can identify the resources required, take appropriate steps to overcome the deficiencies, and provide for future needs.

The test and evaluation needs of the firm depend on the future product needs. These requirements derive from the various future project needs which, in turn, stem from ideas developed informally or formally within the firm. An organized approach to gathering these ideas aids in the thoroughness of defining future requirements. The test and evaluation function must assure that it is in the information-gathering loop so that those requirements are reflected in the total aggregate of future needs. The future requirements of test and evaluation, like all other areas, should be prepared on an incremental basis showing appropriate data such as costs, revenues, and investments to facilitate comparative analysis. Complete management visibility of alternative opportunities is a prime requisite for financial planning and control.

The key element in capital budgeting is the system and techniques for the evaluation, analysis, and dissemination of relevant information. Of primary importance is the establishment of the analytical framework for effective investment proposal or project evaluation.

If the steps previously discussed have been taken, the corporation will have developed project ideas and the relevant information necessary to appraise them. At this point, evaluation criteria must be clearly established in order to evaluate and select the most desirable project or projects. There are several techniques utilized for evaluating and ranking investment projects. They include: (a) payback method, (b) average rate-of-return method, (c) internal rate-of-return method, (d) net present value method, and (e) present value per dollar of invested capital (Table 1).

We recommend (c), (d), and (e). There are serious flaws in (a) and (b), while there are significant advantages to the remainder. Essentially the application of the (c), (d), and (e) methods permits the firm to employ three important concepts: (1) the firm establishes and requires the use of a cost of capital and relates this to each project on the basis of revenues to be received; (2) the time value of money is considered; and (3) the magnitude of the investment is reflected in the evaluation. By evaluating each opportunity and the incremental effects that each method affords, the financial demand upon each project and the firm and the expected profitability for increments of invested capital can be determined. The particular technique employed will depend upon the particular financial policy within the firm. However, it may be advisable to employ more than one technique simultaneously to detect any serious divergence. In the event this should occur, a second look at the data and/or investment opportunity would be warranted. Table 1 summarizes the techniques mentioned above with various pertinent comments included. The reader should explore this area in greater detail and, for this purpose, he should consult the cited volume[3] as one source of information.

The intention of this discussion has been to acquaint the reader with the financial implications of the test and evaluation function, especially when the acquisition of sizable facilities and/or replacements of equipment is contemplated.

EQUIPMENT REPLACEMENT

Now, let us consider equipment replacement. Here we are involved in the purchase of new facilities as well as the disposal of existing facilities. How

Table 1 Investment Evaluation Techniques

Technique	Description	Attributes	Remarks
(a) Payback	Number of years required to return original investment	Simple in concept	Time value of money not considered scope of evaluation Limited to period over which investment is recovered
(b) Average rate of return	Average return over the economic life of project	Simple in concept Extends over the life project	Time value of money not considered
(c) Internal rate of return	Interest rate which equates the present value of future returns to the investment outlay	Considers all revenues Time value of money	Discounted cash flow technique assumes that future reinvestment can be made at the interest rate
(d) Net present value	Present value of future returns discounted at cost of capital minus the present value of the investment outlay	Considers all revenues Time value of money Reflects consideration for capital mix	Discounted cash flow technique assumes future reinvestment can be made at the cost of capital rate
(e) Present value per dollar of of invest capital[a]	Present value of future returns discounted at cost of capital minus the present value of the investment outlay per dollar of invested capital	Considers all revenues Time value of money Reflects consideration for ideal capital mix Reflects the magnitude of investment required via the incremental approach to amount invested	Discounted cash flow technique permits the evaluation on the basis of incremental capital investment

[a]Probably provides the most useful measurement criteria especially when used in conjunction with (c) and (d)

shall we determine when replacement is advisable, and how can we present this information for management review and approval? Here, again, techniques employed by other facets of the operation have meaningful application. One such technique was developed by George Terborgh, Director of Research for Machinery and Allied Products Institute (MAPI), in his *Business Investment Policy* as an approach to decision making in the equipment replacement field. This approach involves the "adverse minimum" of the machines being considered for installation or continuation.

EQUIPMENT REPLACEMENT DECISIONS

The Terborgh approach, usually referred to as MAPI, is simply a method to determine whether the equipment being used currently could be better replaced by a newer equipment. Only two alternatives are considered in the analysis—the machine presently in use (the defender) and the best available new machine (the challenger). The "best" challenger may be selected by the same type of analysis.

Terborgh's concept of the adverse minimum consists of the following two components:

1. *Capital cost.* Determined by the amount of capital investment, the time used, and the firm's cost of capital.
2. *Operating inferiority.* Includes both cost and revenue disadvantages— with improving technology, operating inferiority may be expected to increase with time.

In both cases, the figures must be time-adjusted. This corresponds to a discounting process where the present value of a future dollar is determined. The determination of the adverse minimum of the challenger requires estimates of investment cost and salvage values, of the annual rate of development of operating inferiority [inferiority gradient (g)], and cost of capital.

Table 2 indicates how the adverse minimum for the minimum of the challenger machine would be determined in the no-salvage-value case with the assumptions as given. Obviously, the lowest annual total of capital cost, operating costs, and revenue disadvantages will be realized if the challenger operates for a 12-year period at $1173 per year.

This figure must not be compared with the adverse minimum of the defender machine. It can normally be assumed that the next year's results will be the best future results from the old machine. The problem then becomes the determination of next year's capital costs plus the operating inferiority of the old machine. What is required is the application of the opportunity cost concept to the determination of the capital cost of using the equipment

Table 2 Derivation of Adverse Minimum of a Challenger Having a Cost of $5000 and an Inferiority Gradient of $100 a Year Assuming No Capital Additions and No Salvage Value, with Interest at 10 percent[a] (Dollars)

Years of service	Operating inferiority for year indicated	Present worth factor for year indicated[b] (Col. 1 × Col. 2)	Present worth of operating inferiority for year indicated (Col. 1 × Col. 2)	Present worth of operating inferiority for period ending with year indicated (Col. 3 cumulated)	Capital recovery factor for period ending with year indicated[c]	Time-adjusted annual average for period ending with year indicated Operating inferiority (Col. 4 × Col. 5)	Capital ($5000 × Col. 5)	Both combined (Col. 6 + Col. 7)
	(1)	(2)	(3)	(4)	(5)	(6)	(7)	(8)
1	0	.909	0	0	1.100	0	5500	5500
2	100	.826	83	83	.576	48	2881	2929
3	200	.751	150	233	.402	94	2011	2104
4	300	.683	205	438	.315	138	1577	1716
5	400	.621	248	686	.264	181	1319	1500
6	500	.565	282	968	.230	222	1148	1371
7	600	.513	308	1276	.205	262	1027	1289
8	700	.467	327	1603	.187	300	937	1238
9	800	.424	339	1942	.174	337	868	1205
10	900	.386	347	2289	.163	373	814	1186
11	1000	.351	351	2640	.154	406	770	1176
12	1100	.319	351	2990	.147	439	734	*1173
13	1200	.290	348	3338	.141	470	704	1174
14	1300	.263	342	3680	.136	500	679	1178
15	1400	.239	335	4015	.131	528	657	1185
16	1500	.218	327	4342	.128	555	639	1194
17	1600	.198	317	4658	.125	581	623	1204
18	1700	.180	306	4964	.122	605	610	1215
19	1800	.164	294	5258	.120	629	598	1226
20	1900	.149	283	5541	.117	651	587	1238

[a] Operating interiorities teated as year-end magnitudes. Figures do not always add exactly because of rounding.
[b] The factor gives the present worth of $1 payable at the end of the year indicated.
[c] The factor gives the annuity, payable (at the end of each year) over the period indicated, which has a present worth of $1.

for another year at the firm's cost of capital plus decline in salvage value over the year. To determine the second component of the adverse minimum of the defender requires estimates of all comparative operating cost and revenue net disadvantages of the old machine as compared with the new one next year. These two figures are then added. In this way the adverse minimum of the defender is computed and compared with the figure already computed for the challenger—with the choice between the two made on the basis of the lower adverse minimum.

To avoid the necessity of making the detailed computations of Table 2 in order to arrive at the challenger's adverse minimum, Terborgh has developed various short-cut formulas for use under varying sets of assumptions. These include the no-salvage-value case, which is as follows:

$$\text{adverse minimum} = \sqrt{2cg} + \frac{ic - g}{2}$$

where c = capital investment, g = operating inferiority gradient (the $100/year, in the example, annual rate of development of operating inferiority), and i = capital cost. This formula may also be used to obtain an approximation to the correct adverse minimum except where salvage value or capital additions are very significant factors.

Furthermore, Terborgh states that the life average of operating inferiority and capital cost = [g(n − 1)]/2 + (c − s/n + [1(c+s)]/2

g = the annual inferiority gradient
c = the acquisition cost
s = the terminal salvage value
n = the number of years

This general expression does not yield the challenger's adverse minimum as such, but merely the life average inferiority and capital cost for the particular service life stipulated. To get the adverse minimum it is necessary to attain the service life associated with it.

Let us consider one more hypothetical example. A testing facility considers replacing a 19-year-old vibration machine with a new one. The basic facts are as follows. Initial costs of new machine: $29,860. This year's salvage value for old machine: $6000. Next year's salvage value: $5000. Labor saving due to operating superiority of new machine: $4025. Yearly saving in maintenance: $200. Saving in floor space: $60. On the negative side, we have an increase in taxation and insurance: $230. Given below is next year's operating advantage of the new machine or the operating inferiority of the old machine:

	Next Year's Operating Advantage	
	Challenger	Defender
Direct labor	4025	
Maintenance	200	
Floor space	60	
Property tax and insurance		230
Total	4285	230

Next year's operating inferiority of the defender: $4055
Assuming that next year's operation is optional for the old machines, we obtain the adverse minimum by adding to the above figure the cost of capital of the old machine:

$1000 loss in salvage value
 600 interest at 10% on $6000
$1600 total

Defender's adverse minimum: $4055 + 1600 = $5655. Inferiority gradient is obtained by dividing the operating inferiority by the age of the old machine: $4055 ÷ 19 years = 213.

Using the formula to determine the challenger adverse minimum,

$$\sqrt{2cg} + (ic - g)/2,$$

we solve

$$\sqrt{2(29860)(213)} + \frac{.10(29860) - 213}{2} = \$4952$$

We see that the challenger adverse minimum is sufficiently below that of the defender's to permit recommending replacement.

EQUIPMENT LOADING

As in the field of production control, test equipment loading may be defined in the same manner as machine loading. That is, equipment load is simply the number of hours of work assigned to each machine. It is a measure that changes frequently but requires careful recording to be useful. In addition to the actual running time, the load factor must also reflect the setup or preparation time for each test performed. Since this can far exceed actual running time, records of this phase of the operation are equally (if not more) important. Gantt Charts are frequently used for this purpose.

The objective of load evaluation, as stated earlier, is to obtain optimum output from each piece of equipment. In addition, maintenance of the overall productive capacity of the test and evaluation facilities, synchronized with the end product delivery schedule, is also a primary interest.

The effects of excess capacity were mentioned above; the effects of too little capacity are of equal concern. Quite frequently purchase awards contain incentive clauses for delivery. The test and evaluation program for a new product forms a very significant portion of the development cycle in terms of time. If the facilities are not available or have been improperly scheduled, substantial losses in incentive fee or even penalty payments may be imposed if delivery should be missed.

Several alternate courses of action are available for both situations—over and under capacity. They include subcontracting additional capacity, overtime, improved techniques that may be applied to the test operations for improved operating efficiency. To cite a few, the application of equipment maintenance analysis, detailed scheduling such as program evaluation and review technique, and task analysis have definite contributions to make.

SUMMARY

This chapter provided orientation in basic organization and financial concepts and illustrated these concepts by actual examples. In addition, the examples showed the effects of the nature of the business on the status of the PT&E function. Obviously, in an independent test facility, testing is a primary importance. Economy is also of major concern in an industry where scrap and shutdowns are costly. However, the status of its management, depends on the importance of PT&E in the company. This is an attribute that is difficult to identify and one that varies in the same facility as a function of the individual filling the position.

The emphasis in developing organizations and their cost implications must be on getting jobs done. Historically the emphasis has been on delegating responsibility and authority. Having accomplished the foregoing, means have been developed to prevent abuse. The true essence of organization is to facilitate the accomplishment of the task at hand with minimum constraints but with appropriate check-and-balance control. The conscientious application of the various scientific management analytical techniques to the evaluation task are of significant value in accomplishing this end.

RECOMMENDED READINGS

Cleland, David I., and William R. King. *Systems Analysis and Project Management.* 3rd ed. New York: McGraw-Hill, 1983.

Johnson, Richard A., T. E. Kast, and J. E. Rosenzweig. *The Theory and Management of Systems.* 3rd ed. New York: McGraw-Hill, 1973.

Litterer, Joseph A. *Organizations: Structure and Behavior.* 3rd ed. New York: John Wiley & Sons, 1980.

March, James G., and Herbert A. Simon. *Organizations.* New York: John Wiley & Sons, 1958.

Schoderbek, Peter P., ed. *Management Systems: Conceptual Considerations.* rev. ed., Business Publications, 1980.

White K. K. *Understanding the Company Organization Chart.* American Management Assoc. Research Study No. 56. New York: American Management Assoc., 1963.

Young, Stanley. *Management: A Systems Analysis.* Illinois: Scott Foresman, 1966.

NOTES

[1]Peter F. Drucker, *Managing for Results* (New York: Harper & Row, 1964).

[2]*Modern Product/Operation Managment 7/e.* (New York: John Wiley & Sons, 1983).

[3]*Managerial Finance* by J. Fred Weston and Eugene F. Brigham, 7th ed. (New York: Holt, Rinehart & Winston, 1981).

Chapter 4

Test Facility Applications and Procurement

Personnel alone, of course, cannot accomplish a great deal without the proper PT&E equipment. The capability to implement a meaningful test and evaluation program is also dependent on the facilities provided for that purpose. Management must carefully consider the return on capital investment in this area just as it does for production equipment, since the statement of operations does not differentiate between the two. Test facility requirements are extremely diverse, and a review of facilities serves to illustrate the wide range of requirements stemming from various product evaluation programs being implemented in industry today. A continuing trend (as will become evident in the following discussion) is to develop facilities that accurately represent the use conditions to which the particular product will be exposed. Also of primary interest is the development of test facilities and techniques to enable product evaluation at the system level under operational conditions and to enable the evaluation of equipment interactions and the effects of combined environments. A significant impact on product cost and business profits is the judicious procurement of test equipment, which will be described in this chapter.

The allocation of test and evaluation facilities in a typical industrial concern may be broken down into research and development, production and field installation facilities. Some, or all, will be found in one corporation, depending on the nature of its business. Generally the most specialized (and, hence, the most expensive) facilities are found in research and development. Facilities employed for this purpose involve large-scale environmental simulation. Highly accurate devices used for production testing of products may also

represent a significant investment decision. Some notable facilities presently in operation are discussed to provide an insight into their complexity and diversity. We restrict our discussion to research and production facilities because of the greater use made of them within industry.

RESEARCH AND DEVELOPMENT FACILITIES

A well-run research facility usually has hundreds, if not thousands, of high quality, precision test equipment, permitting experiments in the wide-range fields of, for instance, electronics, frequency propagation, mechanics, guidance and control, thermodynamics, and optics.

The Boeing Commercial Airplane Company's 767-200 Major Fatigue Test Facility was designed to test a 767-200 structurally complete airplane that is representative of production quality in this line of airplanes. It is representative of both a research and production facility operated at the system level of evaluation.

The Boeing 767's structural design, and the new materials and processes used for the first time in an all-new airliner, were proved in the most successful structural fatigue tests ever completed by Boeing.

Using a complete 767 airframe, the fifth built, the fatigue test series took 18 months at the Boeing 747/767 plant in Everett, Washington, and was

---

Figure 1 The equivalent of 40 years of airline flights—twice the normal lifespan of the airline—was carried out by a nonflying Boeing 767, the fifth airframe built, to make certain the new-generation twinjet's structure will be free of metal fatigue problems in service. The airframe test began 18 months ago following extensive preparations—including mounting the wing and fuselage, with vertical fin attached, in a structural steel rig, as shown here—connecting hydraulic actuators to impose the stresses experienced in airline operation, and providing computer control for the tests. Testing has now completed 100,000 "flights," simulating in three and one-half minutes all segments of a one-hour trip beginning with taxi out, through takeoff, climb, cabin pressurization, and cruise, to descent and depressurization, landing, and taxi in. Turbulence and maneuvers have been included in each segment. In separate tests, the horizontal stabilizer, trailing-edge flaps, and landing-gear have also been subjected to fatigue testing. The fatigue tests have resulted in some minor changes to the 767s in production and for 767s already delivered. Service Bulletins will be issued for updating the structure thousands of flight hours before any fatigue damage would come about. Boeing 767 fatigue testing was carried out at the 747/767 plant in Everett, Washington, at the site of the 747 fatigue tests in the early 1970s. After completion of the testing, the fatigue test airframe will undergo a rigorous inspection and then be put into storage for future reference.

completed after 100,000 "flights" — the equivalent of 40 years of airline service (Figure 1).

The test series were described as the most successful fatigue tests in Boeing airliner history because of advanced structural design, materials, and production methods used in the new-generation twinjet. The test series was completed in record time.

Fatigue can occur when metal is flexed repeatedly. Great care is taken in design of airliner structures to prevent such occurrance for the life of the airliner.

Listed as "firsts" established in the 767 test series:

1. The first time Boeing's new Design for Durability methodology has been used in designing an all-new airliner. This made possible concurrent design of structural parts in the U.S. and two foreign countries to the same rigorous standards.
2. First double lifetime proof of fatigue properties of new light-weight, high-strength aluminum alloys developed by Boeing and the Aluminum Company of America and used extensively for the first time in the 767.
3. First fatigue test proof of the continued durability of the largest graphite composite structures in airliner use.
4. First double lifetime proof of new Briles rivets' excellent fatigue reducing characteristics. Briles rivets were used extensively for the first time in an airliner on the 767.

The first lifetime, fatigue testing was intended to find any discrepancies so they could be corrected and corrections put into 767s in production and in service long before a 767 in the airline fleet service accumulated as many flight hours. The fatigue testing continued into the second 767 lifetime with the additional aim of determining where, beyond the normal airliner lifetime, eventual fatigue cracking could occur. These data are used to validate the airline structural inspection program.

As it turned out, design changes resulting from the fatigue testing were minor and the testing continued straight through to the end of the second "lifetime," another Boeing first.

The detailed structural design of the 767 was carried out by about 1000 engineers at Boeing plants in the Seattle area, and in Wichita, Kansas; by Aeritalia in two Italian cities; and by the Civil Aircraft Corporation (formerly Civil Transport Development Corporation) in three Japanese locations. Production was then carried out to Boeing specifications and standards. Major structural parts also were produced by Canadair in Montreal, Canada; Convair in San Diego, California; and Grumman Aerospace Corp. in Stuart, Florida.

Preparations for the fatigue test series began two full years before the fatigue test airframe was put in place in the massive structural steel test rig at Everett. Hydraulic actuators were installed to impose the stresses experienced in airline operation, and computer control and recording was finalized.

Testing consisted of 100,000 one-hour "flights"—simulating in three and one-half minutes all segments of a typical trip, beginning with taxi out through takeoff, climb, cabin pressurization, and cruising flight, to descent and depressurization, landing, and taxi in. Turbulence and maneuvers were included in each "flight." In separate tests, the horizontal stabilizer, trailing-edge flaps, and landing gear were subjected to equally rigorous testing.

After completion of the testing, the fatigue test airframe will undergo a full inspection and then be put in storage for future reference.

PRODUCTION FACILITIES

The production phase of the producer-to-user cycle represents an extensive amount of product evaluation activity. As noted earlier in the discussion of the economics of product evaluation, the amount of activity as represented by manufacturers' expenditures is significant. It is likely to increase in importance and cost in the future. Although production testing does not always involve environmental simulation, it is an important aspect of the systems test philosophy. In addition, major considerations in almost every instance are testing speeds, accuracy, and comprehensiveness, especially in mass production, process, and high technology industries. For these reasons, production test facilities represent significant assets and pose challenging technical problems.

Some noteworthy production testing techniques and facilities representing these ideas are discussed below.

Conventional radiography has historically been employed as one of the basic nondestructive test or inspection techniques. The traditional approach, based on radiographic film technology, has some inherent limiting factors that can impact overall productivity in the modern factory environment. For example, the time required to align the part for the radiograph and to develop the film can introduce significant flow-time factors: particularly on complex and multiple exposure evaluations. The cost of the film can impose limitations on the number of exposures taken, hence, degree of evaluation of the product. Film storage also involves significant costs in high volume activities where historical records must be maintained for an extended period of time. And, all of these activities are labor intensive.

The Boeing Aerospace Company has developed a high resolution nonfilm radiography system with automated robotic parts handling and remote manual or programmed parts manipulation in a four-axis flexible motion envelope.

Figure 2 is a conceptual block diagram of the system and Figure 3 describes a typical installation in an automated factory radiographic vault.

The features of this system include:

1. Radiographic sensitivity up to 2% resolution.
2. Remote automated robotic parts handling in a radiographic environment.
3. Flexible, real-time, four-axis part manipulation and radiographic display.
4. A five-to-one reduction in inspection and inspection material, equipment and storage costs.

Typical applications include evaluation of production welds and procured castings and real-time inspection of microelectronic parts and assemblies as well as completed mechanical assemblies.

The application of traditional dimensional measurement inspection techniques in a modern automated mechanical factory would be a significant impediment to overall factory productivity. The traditional surface plate, height gage, manual layout, etc., approach would have capacity to handle the high volume production output as well as insufficient inspection accuracy to properly evaluate the high precision and complex part geometry capabilities of a modern mechanical factory.

The Boeing Aerospace Company has also developed a computer-controlled inspection cell combining robotics for material handling with automated inspection techniques for dimensional verification (see Figures 4 and 5). The system consists of a computer-operated robotic material handling system which loads and unloads two automated coordinate measuring machines and a robotics inspection station in the inspection cell. A central "host" computer monitors and coordinates these functions, optimizing utilization of both inspection machines and the material handling facility. Special part holding fixtures and handling pallets have been designed for close tolerance locating, hardware indexing, and inspection point access. The automated part inspection programs are developed directly from the computer-aided design and manufacturing data. Collection and analysis of inspection results is automated, including accept/reject decisions, inspection report generation, and severity adjustment. Anticipated time and cost savings improvements over the traditional manual methods are 50 to 60%.

Figure 6 is a partial photograph of this facility which is designated the Flexible Inspection System (FIS). The system is designed for co-location and integration with the Boeing Aerospace Company's Flexible Machining System (FMS).

AT&T's North Carolina Works manufactures an assortment of wired panels, backplanes, and bays that provide the interconnection between the printed wiring boards in a system as well as the interconnection between systems.[1] Before shipment, the wiring of every product must be verified. Of

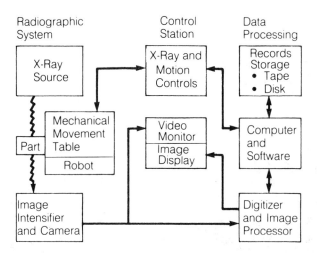

Figure 2 High resolution nonfilm radiography system conceptual block diagram.

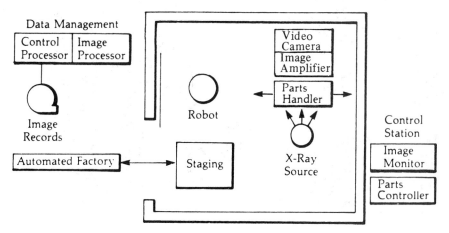

Figure 3 Typical system installation in an automated factory radiographic vault.

these products Bellpac backplanes present a challenging test problem, primarily because of their high pin density. However, the advent and acceptance of the Bellpac equipment standard has prompted the development of test set hardware and software that can be used across current and future product lines.

Backplane interconnections may consist of wire, as with the conventional backplane, or they may be a combination of wires and printed circuit packs, as with the Bellpac backplane shown in Figure 7.

The Testing Challenge

With respect to wire verification testing, Bellpac backplanes differ from conventional backplanes in three significant areas:

1. Connection from the Bellpac backplane to the circuit packs is by .025-inch square pins spaced on a .125-inch grid rather than by conventional connectors.
2. The Bellpac backplane pin density—up to 64 pins per square inch—is higher than conventional backplane pin densities.
3. Guidelines exist that specify the sizes and locations of circuit packs in the Bellpac backplanes.

To solve these problems, a system consisting of hardware and software was developed for wire verification of Bellpac backplanes (Figure 8). A zero insertion force connector replaces the standard female connector and remote solid-state switch modules that mount on the backplanes to be tested replace the centralized relay switches. Instead of "one wire per point" to a centralized cabinet scheme, a bus controls the remote switches, resulting in fewer cables and a more manageable setup. A comprehensive package of test generation software was also developed to support the new test system.

Test Set Hardware

The remote switch module is at the heart of the Bellpac equipment test set. It consists of four major components:

1. Zero insertion force connector
2. Switchboard (Figure 9)
3. Control board
4. Control bus

Figure 4 One of the automated inspection stations in the Boeing Aerospace Flexible Inspection System (FIS) is a robotized utrasonic thickness measuring station. Here the station measures the skin thickness of an air launched cruise missile wing. The robotized materials handling vehicle is in the background.

Figure 5 A close-up of the robotized ultrasonic thickness measuring station in the Flexible Inspection System.

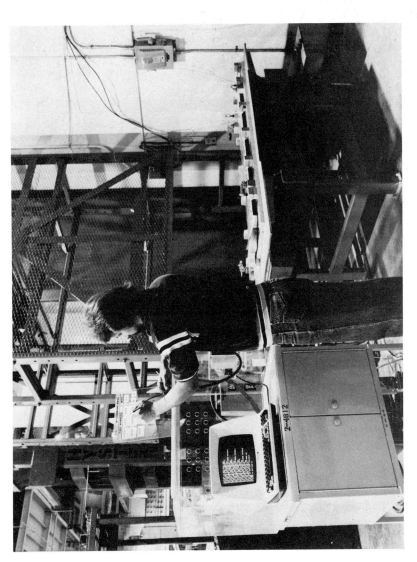

Figure 6 A load/unload station for the Boeing Aerospace Flexible Inspection System (FIS). A worker loads a part on a pallet at the station and communicates with the host computer using an electronic wand and bar codes at the terminal. The robotized Hyster forklift in the background serves as the materials handling link for the FIS, automatically moving materials into and out from the individual automated inspection stations.

Figure 7 This backplane is part of the Remote Trunk Test Unit/Central Trunk Test Unit system, which automatically and periodically tests trunks and locates faults in many different switching offices. Each wired connection must be verified before the unit is shipped, a task which becomes challenging and complex as the number of interconnecting points increases.

Bellpac equipment guidelines allow for three different circuit pack widths, commonly referred to as eight-inch, six-inch, and four-inch;[2] therefore, the module is available in three sizes. The switch module is enclosed in an aluminum housing which can be plugged into the backplane shelf.

The Test Set Software

Backplane testing requires that many tests be performed between all the terminals on the backplane. Consequently, the test programs and other data files that the test engineer creates and uses are often large. Software running on the Unix* operating system assists the test engineer in program development. The test set controller contains the necessary software to read and execute these programs, display test results, and interface with the operator.

*Trademark of AT&T Bell Laboratories.

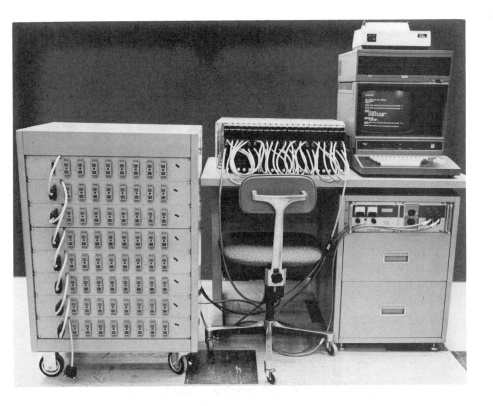

Figure 8 The test set hardware for wire verification testing of Bellpac backplanes consists of, from left to right, a 3648 point miscellaneous connector switch unit, remote switch modules, power supplies, and the system controller.

Progressive assembly, conveyorization, and the latest soldering technology are used to increase productivity in high running circuit packs at AT&T's Denver Works (Figure 10).[3] The new "fast line" process reduces handling, process time, and floor space, and improves overall product quality.

The fast line, operating on a five-day, two-shift basis, can produce 15,000 circuit packs a week. It occupies only 9500 square feet of floor space—about 35% less than the original batch processing shop, which was essentially a manual operation—and it requires just 40 operating personnel instead of the 70 previously needed.

The line has five sections: board preparation and machine insertion; progressive assembly; wave soldering; detergent cleaning; miscellaneous as-

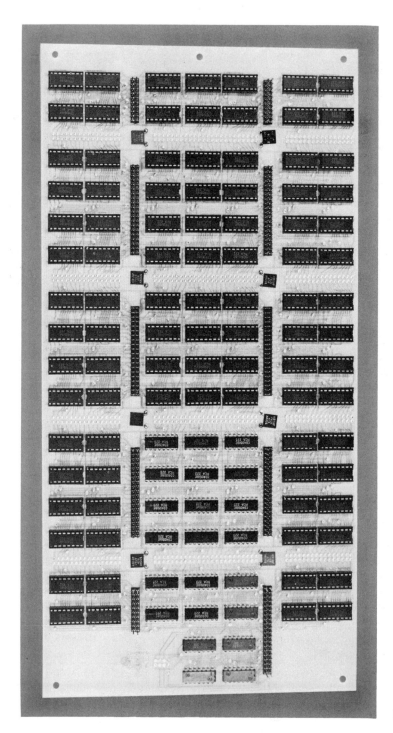

Figure 9 This eight-inch switchboard contains 116 CMOS integrated circuits and measures 7.5 inches by 15 inches. Because of its modular design, one section can be eliminated to make the six-inch switch, and two sections can be eliminated to make the four-inch switch.

Figure 10 Progressive assembly, shown here, is one of five sections that make up the Denver Works' "fast line" for assembling and soldering circuit packs. In this section, operators manually insert those board components that cannot be handled by machine.

sembly, touchup, and process check. All but the first section are equipped with conveyors that automatically transfer product from one step to the next.

Board Preparation and Machine Insertion

At this initial stage, all incoming material is checked and prepared for assembly. This includes the printed wiring boards and electronic components from which the circuit pack assemblies will be built. The boards are inspected for major construction flaws and for any defects in the gold finger connectors that might affect electrical contact. Some components are actually assembled into printed wiring boards at this stage, as in the case of three-legged transistors. The legs of these devices must be clinched so that they can be soldered properly. This requires access to the noncomponent side of the circuit pack, which would not be possible during progressive assembly. Also in the board preparation area, integrated circuits are prepared for progressive hand assembly by being processed through a trimming and preforming device.

Following board preparation, the circuit packs go to the machine insertion area (Figure 11). Here, dual in-line packaged integrated circuits and devices with axial leads are inserted by high-speed, computer-controlled machines. Four dual-headed machines for axial lead components and one for dual in-line packaged circuits are sufficient to handle production volume. One dual-headed machine is used solely for standing some components off the boards to allow for heat dissipation or to provide adequate clearance for the subsequent aqueous cleaning process.

Machine insertion is not always warranted. Sometimes components are too sensitive to trust to automatic handling or, as in the case of hybrid integrated circuits, too expensive. Often the quantities per board are too small to justify the cost of automatic insertion, given the set-up time involved. In these cases, components are inserted by hand on the progressive assembly line.

Progressive Assembly

This section of the fast line handles components that must be inserted manually such as relays, transformers, light emitting diodes and hybrid integrated circuits. All but the integrated circuits are trimmed and preformed on line.

The section consists of a motorized conveyor and ten assembly positions. At the first position, an operator takes a machine-inserted circuit board from a truck, places a small profile conductive boot over the gold fingers, and loads the circuit pack onto the conveyor.

As the conveyor carries the boards past successive positions, an operator selects a component in each hand from a gravity feed dispenser, inserts the components in a pneumatic lead cutter if trimming is required, and places the components in the board (Figure 12). Each operator places approximately four or more components in the board in this manner, depending on the code

Figure 11 The first step in the fast line process involves machine insertion of components into circuit boards. Dual in-line packaged integrated circuits and devices with axial leads are placed in the boards by high speed, computer-controlled machines.

Figure 12 A gravity feed dispenser supplies components to an operator in the progressive assembly section. Where necessary, the components are trimmed in the pneumatic lead cutter located in front of the operator before being inserted in the circuit boards.

being run. The last position is used as a process check position to ensure that components are properly oriented and seated.

Commercial gravity feed dispensers are used to supply integrated circuits to the assembly operators. For bulk components, dispensing trays angled toward the operators perform the same function. The trays, moreover, are sized to accommodate direct transfer of components from packing material. The high volume of the line requires that the setup of each position provide enough components to last four hours.

Progressive assembly offers several advantages over batch processing. One is that queuing between operations is minimized. This alone has reduced work in process for presolder assembly by 80% and represents a substantial reduction in investment. This also means that floor space requirements for product staging have been decreased. Additionally, minimized material handling has resulted in lower labor costs. And finally, the continuous flow of work stimulates operators to work at a consistent, high rate of efficiency.

Wave Soldering

The wave soldering process used on the fast line involves fluxing, preheating, and soldering. The conveyor speed is set at five feet per minute. At this rate, the best solder quality is obtained, and it is this speed that paces the entire line (Figure 13).

Detergent Cleaning

A new approach was tried in cleaning flux residues from the printed wiring board assemblies. In wave soldering, AT&T has customarily employed an underbrush process using chlorinated solvents, such as perchloroethylene and trichloroethylene. However, growing concern over the environmental effect of chlorinated solvents, coupled with the excellent cleaning ability of detergents, prompted an investigation into the possibility of using detergent cleaning on the fast line.

Detergent cleaning of rosin flux is a threefold process. First the rosin is saponified by washing the boards in a detergent solution. Next the boards are sprayed with water from top and bottom to remove the rosin soap and any residual detergent. Then the boards are thoroughly dried by radiant and convective heating.

There are three methods of checking to ensure that the boards have been thoroughly cleaned. The simplest is a 100% visual inspection for residues. A second check, performed approximately every two months, measures the percent of detergent remaining in the wash solution by titrating a sample. The titration value corresponds to a specific concentration of detergent as established empirically under controlled conditions. A minimum detergent concentration of 4% generally provides adequate cleaning. (The minimum

for lines at other locations will vary according to production load, type of solder mask used, and type of flux employed.)

A third check, which is the main process control method, is conducted twice per shift by measuring the resistivity of a solution of alcohol and deionized water after a "cleaned" board has been immersed in it for a set time. Resistivity is expressed in units of megaohm-cm. This can be converted to equivalent units of milligrams of sodium chloride per square inch (mg NaCl/in.^{2}), which is a common way of expressing cleanliness of circuit packs. Since there were no Bell System requirements for ionic contamination of printed wiring board assemblies (as indicated by resistance levels), the military specification (MIL-P-28809) for cleanliness was used as a reference.

Miscellaneous Assembly

The last conveyor in the fast line feeds the circuit packs to the miscellaneous assembly, touchup, and process check positions (Figure 14). Here, operators add whatever components could not be inserted earlier, such as those that would not lend themselves to detergent cleaning. They also perform touch-up soldering and inspect the overall product.

The final product evaluation facility used in production to be discussed is the functional test setup used to apply functional tests to each of the 256,000-bit dynamic random access memories manufactured at AT&T's Allentown Works (Figure 15).[4]

The chip being probed is integral with several hundred others on a wafer. If it fails any test, it would be culled out of the good ones that go on to electronic packaging. The test system used is also a process control tool. Software was developed which includes "intelligent" ordering of tests and "intelligent" decisions to abort testing after certain failures. Consequently, analysis of the test data can lead a product engineer to the source of the trouble. Testing of large capacity memories can be time consuming and costly. Trade-offs must be made between test efficiency and the necessary process control feedback. Much must still be learned about testing a memory chip like this to produce a less expensive but highly reliable product.

PROCUREMENT OF TEST EQUIPMENT

Readers know the significant impact that a properly implemented evaluation program can have on product development and improvement. They also

---

Figure 13   Partially completed circuit boards move by conveyor into the wave soldering machine at the rate of five feet per minute. This is considered the optimum speed consistent with good solder quality for these assemblies and sets the pace for the rest of the line.

Figure 14 Miscellaneous assembly and touchup are performed at the end of the fast line. Here, final insertion of components is made and the product is inspected for overall quality.

know that in business enterprises the real basis of a choice between alternatives is the prospective effect of each alternative on the costs and revenues of the business. What they may not realize is that the selection of test equipment for PT&E carries a significant impact on product cost and business profit. This section describes a modern approach to the procurement of complex test equipment and, more specifically, state-of-the-art equipment, applicable to both software and hardware systems.

Emphasis is placed on the system concept, rather than on the discreet component approach to modern equipment. As equipment becomes more complex, it becomes increasingly difficult for the PT&E engineer to know the detailed operation of each subassembly. Not only that, but demands for getting a job done do not permit him enough time to educate himself in the working details of equipment to enable him to assemble a complete system from the basic components. His job is to perform tests, gather and evaluate data, and not to build equipment.

There are situations where a desired piece of equipment is not available and must be developed. In such a case, it may be more economical to have the equipment manufacturer develop this item rather than to enter an unfamiliar environment. Equipment manufacturers, well aware of this situation, are becoming more system oriented. In short, manufacturers are equipped to supply complete systems and assume system responsibility. This relieves the PT&E engineer from the task of selecting components or units that are compatible with each other, and of coordinating purchases from a number of suppliers. However, he is responsible for the decision of selecting the supplier of this test or data system. This responsibility means more than simply selecting the supplier who is a low bidder on a request for quotation, even though the response is to a detailed specification. The following example and detailed suggestions give the PT&E engineer a set of guidelines for taking this approach to purchasing of a system, rather than an accumulation of test instruments and recorders which will, hopefully, function as a PT&E system.

The situation often develops in this manner: you, the PT&E engineer, or laboratory manager, are requested by management to improve your organization's overall capability in a certain area of PT&E, or you are requested to procure equipment to meet the requirements of a specific program. This is done for various reasons: to improve your organization's overall competitive position, to meet the increasing demands of those people or organizations that you service, and to perform newer types of tests as they develop. The overall span for such procurements can be reduced by initiating the following program and procedure for specifying and procuring a complex system.

Responsibility

One individual must assume the responsibility to manage and implement this project, since such a project may involve many people and organizations that may be physically separated from one another. The end users of the PT&E equipment may be many, and if a supplier attempts to satisfy all of their individual needs through separate communication channels, it is very likely that the system will be either unnecessarily expensive or so complex that the desired task may be difficult to perform. An assigned individual makes it much easier for all parties involved to communicate and coordinate.

Time Schedule

A schedule must be set up. It should include sufficient time for potential suppliers to respond to the request punctually, and should include visits to installations that have similar equipment. This schedule depends on the complexity of the system, and cost of the system, and its state of development. Travel to similar installations is usually not necessary unless the system is fairly large, complex, or comprises state-of-the-art equipment.

Equipment Search

There are a number of good sources of references for recommendations of equipment suppliers. Unless the equipment is standard, it would be wise to consult all of the sources mentioned below. Even if you think that you know most suppliers, you may find better ones. A check list of the referenced sources should include: (a) testing laboratories, (b) trade magazines, (c) technical journals, (d) universities, (e) personal contacts, and (f) trade shows and technical society meetings.

Other testing laboratories may already have made a similar purchase, or may have had contacts with potential suppliers. When contacting these organizations, you should inquire about the capability and reliability of the potential suppliers in your general area of interest. If you tend to be a "do-it-yourselfer," it would be wise to solicit opinions on this approach to modern equipment. Many PT&E engineers, who have assembled systems from components and have struggled through problems, are strong advocates of the purchased system approach.

Trade magazines, supported by their advertising, are usually supplied free of charge to qualified personnel. Even if you have already read the articles in these magazines, you should go through them again for the sole purpose of locating desired equipment suppliers.

Technical journals not only include advertising by potential suppliers but also new equipment releases that may be of interest where procurement is oriented toward new equipment.

Universities are also a good reference source. Professors and graduate students in a particular field of interest are often in contact with potential suppliers through their technical society associations. Suppliers often seek them out, since many graduate projects result in new testing philosophies and/or new test equipment.

Personal contacts include business associates and members of technical societies. One useful contact that many people frequently overlook is sales representatives, especially those selling associated products. As an example, a representative for a strain gage or load cell manufacturer may also represent a supplier of testing and data collecting machines, or may know of some manufacturers through his own customers. Most sales representatives of highly technical products are strongly service oriented and are normally eager to assist a potential customer.

Many technical societies stage equipment exhibitions at their annual meetings. Participating exhibitors are, of course, eager for you to visit their exhibits, which are increasingly manned by technically strong personnel. Information and technical literature are free and plentiful at these exhibits.

Communications

Once you have accumulated pertinent technical literature and have talked with the supplier's technical representatives, you will probably be confronted with a number of unfamiliar terms and expressions. You may also encounter an entirely new idea or method of testing and new data processing equipment. It is not necessary to become an expert in equipment design to be able to understand the operating principles and capabilities. It would be advantageous, however, to have a glossary of terms commonly used by the equipment manufacturers. Most suppliers are more than willing to hold brief seminars for you and your staff to acquaint you with the principles of operation and applications of their equipment. They are well aware of the value of having their technical personnel present papers to technical societies or to give seminars. They will therefore probably have a number of technical publications describing their equipment and its applications.

Preliminary Investigation into Capability and Cost

Before progressing deep into this project, the cost of desired equipment must be weighed against the allowable budget (if one has been established). If a budget has not been established, at least "ball park figures" must be arrived at in order to allocate the proper funding. When tentative cost figures vary considerably from the allowable budget, you will then either have to alter the equipment requirements or solicit additional funds. The actual placement in order of sequence of this step will vary, depending on the particular application. You should discuss the subject early enough so your time or the equipment supplier's time will not be wasted. Any request should include enough flexibility to permit selection of various alternatives and options in order to stay within a budget.

Some of the most common mistakes or cost item omissions are: (a) shipping costs, (b) installation costs, including installation of proper utilities, (c) travel costs for training and checkout purposes, (d) taxes, and (e) duty and brokerage fees where applicable. These items represent a significant percentage of the entire system cost.

Preliminary Proposals or Quotations

It is advantageous to have potential suppliers submit preliminary proposals, describing a system that they feel will meet your present and future requirements. In this age of rapidly changing technology, it is advisable to have a system that offers a degree of flexibility so that, as requirements change, the equipment (with minor modification) can adapt to the new demands. These preliminary proposals should include a number of alternatives or options in order to balance capability against budget.

There are situations, as in the case of a system that is modular in concept (a complete system comprised of a number of subassemblies or modules), where a complete price breakdown may be advisable. This breakdown is often difficult for the system supplier, since the cost of each subassembly or module is determined by the particular application, interface requirements, and associated systems engineering. In all cases there is a wide degree of flexibility in choice of equipment and capability of a modular approach to a test system.

Detailed Equipment and Overall Performance Specification

Some suppliers will assist with the details of such a specification. A word of caution: taking what appears to be the best or tightest specification from a number of suppliers may result in all suppliers having to raise their prices considerably. It is best to be realistic in outlining the details of the specification and not to make some items so tight as to cause an unnecessary increase in cost. Here again, potential suppliers will tell what can be readily accomplished and what the relative cost for increased performance is. Another word of caution: ask suppliers who produce an extremely tight performance specification to show how they arrive at their figures and to show actual test results. Any preliminary specification should be sent to suppliers for their comments before it is submitted as the final form. Request firm proposals for quotations from qualified bidders based on the specification.

Evaluate Final Proposals

Regardless of whether you must accept the low bid, you must evaluate all bids. Some low bidders may disqualify themselves on one or more of the following points: experience, reputation, service, ability to assume full system responsibility, company stability, and engineering capability.

Your primary responsibility as a PT&E engineer is to accomplish a complete program, not to sponsor the research and development of a test system. In either case, consider only those suppliers who have had similar experience and can give you references. All bidders should be requested to give a list of customer references who have similar equipment. By all means, contact these references.

Depending on the complexity of the system, it may be desirable to be able to obtain service readily from the supplier or his local representative. As systems become more complex, the maintenance of a staff of qualified technicians becomes more difficult. The increased use of solid-state circuitry and the increased reliability of mechanical components enhance the overall system reliability and thereby hopefully lessen the need for service. Most system suppliers maintain a staff of highly qualified service personnel who are available. Service includes routine updating of equipment, training of personnel, and overall response to inquiries.

SUMMARY

The importance of the proper evaluation, procurement, and operation of PT&E equipment has been discussed. Case histories were given and may serve as usable criteria for the reader's applications.

NOTES

[1]Adapted from Robert D. Bridges, Holder, K. Andrew, Price, Donnie M., "Wire Verification of Bellpac Backplanes," *The Engineer*, Third Issue (AT&T Technologies, 1983), pages 49-55.

[2]The true circuit pack widths are 7.67 inches, 5.67 inches, and 3.67 inches.

[3]Adapted from Cheryl L. Jones, Plomondon, James L., Hanson, Conway M., "New Fast-Line Process Improves Productivity in Circuit Pack Assembly," *The Engineer*, First Issue (AT&T Technologies, 1983), pages 57-61.

[4]Back Cover, *The Engineer*, Second Issue (AT&T Technologies, 1983), page 98.

Chapter 5

Product Evaluation Information Systems

This facet of the PT&E function represents a most important part of the overall product evaluation program. Product decisions will be no better than the data they are based on. Therefore, much planning must be given to the information system when it is being established. Too little information is almost useless while, on the other hand, we can create a monster of paperwork, which becomes difficult if not impossible to control and use.

An important concept to remember is that the data and the system for handling it are the means of providing engineering and management with timely, valid information upon which to base decisions. Data requirements, collection, analysis, and reporting to the decision maker must reflect satisfaction of that concept. This chapter presents each aspect of the information system, with examples to illustrate its application.

DEVELOPING THE DATA COLLECTION SYSTEM

A primary consideration in establishing any information system is to identify what type of data are meaningful to a particular situation. In addition, the methods of analyses require predetermination. Carefully considered answers to these factors will provide invaluable guidance in developing the information system.

Product evaluation data may be collected for a variety of reasons: product research and development, process controls, product improvement, quality control, and marketing intelligence. Additional reasons not quite as obvious, but of increasing importance, are product liability suits and recalls.

Manufacturers can reduce liability losses by keeping detailed records of the reasons for acceptance or rejection of product designs, the testing and inspection procedures that were followed, and the packaging and shipping methods that were used. Then, court cases are viewed more favorably because the company has proved itself careful in the design, manufacture, and shipping of products. In addition to manufacturer-generated data, data must also be collected from the field. Proper notation of the use conditions by sales engineering personnel with signed customer acknowledgment of the pertinent facts, along with failure-analysis data, should be thoroughly investigated and filed. The need to recall items from the production line or the field adds an additional data collection requirement: traceability. Newspapers attest to the importance of this information, citing again and again the recall of vehicles and other products by their manufacturers. Automobile and tire owners now keep manufacturers informed of their addresses to aid in recall programs.

Complex products and highly variable products normally require the collection of significant quantities of variable and attribute data. For instance, paper is not as uniform in structure as are many other products such as plastic, glass, and steel; its properties vary slightly from sheet to sheet and from place to place within a sheet. These basic properties and product characteristics of paper are considered when making tests. They must also be considered when structuring the data collection system in connection with these tests.

The final determination of what—and how much—data to collect should be governed by the cost and difficulty of gathering it and the risk involved in not doing so. This situation lends itself to the concepts of statistical decision making.

STATISTICAL DECISION THEORY

The decision problem on data collection is what and how much to collect. The answer to what data we need will depend upon what we want to know about the product. What is the parameter of interest? If it is service life, life data must be gathered; if it is strength, then breaking-strength data are necessary. The other aspect is to determine how much data on the selected parameter we need to collect.

This decision problem can be approached, generally, by identifying the components of the problems as follows:

1. A set of two or more strategies or courses of action, one (and only one) of which must be selected by the decision maker, such as shutting down the line, or stopping shipment of the product, or the testing of 0, 1, 2, 3 . . . specimen.
2. A set of possible outcome states or states of nature, one of which occurs

at some future time or has occured and is not yet known, such as variation in the parameter of interest.

3. A set of consequences or payoffs to the decision maker, one of which corresponds to every *pair* of strategy and outcome state situations in which the decision maker may find himself after making a strategy selection and experiencing an outcome state. For example, the possibility of stopping the line and finding nothing wrong would yield a certain cost to be deducted from profits.

4. A set of objectives against which the possible consequences are compared in the selection of a strategy—for example, maximum profit or lowest cost.

5. A set of criteria by which the decision maker relates his objectives to consequences and ultimately selects his course of action, by choosing the strategy with the largest "expected" payoff. Table 1 shows the foregoing, diagrammatically.

In Table 1 the parameter states are possible values of the parameter that we are interested in. Criterion of choice, the expected payoffs [E(P)], are the values associated with each alternative (A) given the resultant consequences (C) for each parameter state(s). This could, for example, be the number of defective units or the average value of some parameter, such as weight, length, or strength.

As an example in which the consequences are known quite precisely, consider the situation where the firm produces a product to certain specifications. The product parameter of interest varies, resulting in a normal distribution.

In its primary decision problem, the firm must decide among the alternatives whether to ship without inspection (A_1), test and ship if the product is acceptable, rework and then ship if product is unacceptable (A_2), or scrap everything if the product is unacceptable (A_m). The profits vary for each alter-

Table 1 Basic Components of a Decision Problem

Alternatives	Parameter states(S_i)				Expected payoffs E(P)
	S_1	S_2	S_3 ... S_N		
A_1	C_{11}	C_{12}	C_{13} ... C_{1N}		$E(P_1)$
A_2	C_{21}	C_{22}	C_{23} ... C_{2N}		$E(P_2)$
.	.	.	.		.
.	.	.	.		.
.	.	.	.		.
.	.	.	.		.
A_M	C_{M1}	C_{M2}	C_{M3}	C_{MN}	$E(P_N)$
		Consequences			

Table 2

Parameter state (S_i)		Payoff (P)			
S_1	Alternative	A_1	A_2	. . .	A_M
S_1		$P_{A_1S_1}$	$P_{A_2S_1}$		$P_{A_MS_1}$
S_2		.			
S_3		.			
S_N		.			$P_{A_MS_N}$

native and depend on the existing but unknown value of the parameter of interest.

The first alternative (A_1) payoff yields a profit determined by the difference between the selling price and manufacturing cost less the cost of customer returns due to defects that would have been caught by inspection. Alternative (A_2) payoff must reflect the cost of test and rework of defectives. Alternative (A_m) yields a profit reflecting the costs due to testing and scrap.

Now, slightly modifying our previous diagram, we prepare a matrix (Table 2) that reflects the payoff for each combination of parameter state and alternative where decision making under certainty exists. If a census were taken to determine the parameter states as illustrated, the payoff data can be reflected directly into the table. However, if sampling techniques are employed, the results are subject to sampling errors. To cope with this event, the probabilities associated with each alternative for various assumed values of parameter states relative to a predetermined decision rule must be established. By a decision rule, we mean for example: (1) census product and determine S_n, and decide accordingly, or (2) take a simple random sample and calculate the average and if $S_i = S_1$ do A_1 or if $S_i = S_2$ do A_2 and so forth.

Having accomplished the foregoing, the next step is to prepare a table (Table 3) of the payoff calculations for the various decision rules where decision making under certainty exists. E(P) is simply the product of the payoff value and the expected frequency of occurrence or probability of the parameter states S_i. The matrix, when completed, will reflect the payoff for each act and depends on the true parameter value. In this case we are treating the parameter value as unkown and uncertain; that is, we have no idea as to the likelihood of whether one value of S_i or another might be true. Each act can then be treated as an alternative whose payoff is uncertain, depending on the parameter state. Criteria for decision making under uncertainty therefore apply. For example, under maximum profit, the minimum profit point is found for each act, and the act yielding the maximum of these minima selected leads to a conservative decision.

Table 3 E(P) Calculation

Parameter state (S$_i$)	Alternative	Payoff calculation E(P) A_1	A_2	...	A_M
S$_1$		$E(P_{A_1S_1})$	$E(P_{A_1S_1})$		$E(P_{A_MS_1})$
S$_2$		.	.		.
S$_3$		.	.		.
S$_N$		$E(P_{A_1S_N})$			$E(P_{A_MS_N})$

The maximum decision would be the act leading to the maximum payoff. One can also apply the concept of opportunity costs to decision making. One can utilize the criterion of regret by subtracting the largest payoffs from each of the other payoffs and then selecting the act with the minimum of the resulting values.

Thus, by applying the criterion for decision making under uncertainty which harmonizes best with the decision maker's attitude, an appropriate act is selected. The act and decision rule provide a solution to the original decision problem, since they indicate which course of action to take in the light of future expectations.

PRODUCT CONSIDERATIONS

Data collection for a process industry product (such as paper) often becomes an integral part of the production process. Test data are recorded on charts that provide a graphical presentation of whether a specific property is within control limits. Data are converted to a form for a computer application, which can then be automatically and statistically analyzed for conformance with specifications and for observation of trends.

A complex product, consisting of an assemblage of parts and subsystems, frequently involves the use of a logbook which accompanies the equipment through the assembly and test sequence. The logbook, in effect, provides a complete history of the final end item, including a record of all test data and inspections performed. If the data are to be analyzed automatically, a data recording form (suitable for this purpose) should be used.

The use of specially designed forms facilitates the data-handling process by providing a comprehensive document on which to record the data. Properly conceived, it can facilitate automatic analysis and can become an integral part of the final test report.

As was mentioned earlier, the need for automatic recording of data is essential where the quantity of data is great or the speed of acquisition is high.

As an outgrowth of this need, many automatic equipments have been developed. In brief, the equipment can establish a specific test condition by providing a series of stimuli to the item under test as directed by a computer input. Instruction codes also provide for measurements and automatic recording. The flow diagram of Figure 1 illustrates the use of this equipment in an integrated recording and analysis system.

The principal advantages of such a system are that it eliminates the need for much tedious manual data recording, reliable error-free data recording, and performance "in-line" comparative analysis, thereby providing "real time" feedback to the monitor.

ANALYSIS

The nature of analysis is the next step in the data processing system, and depends on the type of data involved. Large quantities of variables data offer an excellent opportunity to perform sophisticated statistical analyses. On the other hand, customer data feedback may not be conducive to statistical analysis at all.

Data analysis is performed for two purposes: (1) to determine the status of a given product or process relative to pre-established criteria, and (2) to establish what the product or process capability actually is. Since statistical techniques for both are well known, they are not discussed here. Notice, however, that prior consideration of the statistical techniques to be employed will provide the most efficient use of test specimen and data.

Statistical experimentation differs from standard day-to-day methods of testing by varying more than one factor at a time to determine its effect. The fundamental basis of statistical experimentation is to apply the factors in such a way that their effects can be mathematically factored out of the design.

The variations caused by the factors' effects can then be measured and compared with each other and against random variation. A clearer picture of the types of analysis involved may be obtained by considering the following examples which illustrate the efficient use of test data.

Example: Gas Turbine Blades

Gas turbine blades were scrapped for thin walls after rework for spalling. This represented a high-cost scrap item, but 15% of the original blades required rework. The following case history is presented in outline form to illustrate the economic advantages of data analysis techniques.

Problem

Blades were scrapped for thin wall after rework for spalling. Not only was this a high-cost scrap item but 15% of the original blades required rework.

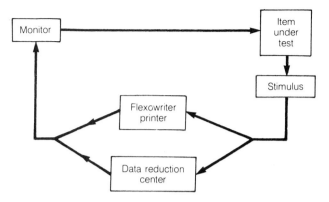

Figure 1   Typical data-flow diagram.

Source of Information

Scrap Reports from accounting and hold-order analysis of scrapped items.

Investigation

Analysis revealed 40% of the reworked blades were being scrapped.

Step 1

Since stripping prior to rework was suspected as the cause, two production lots were sampled (26 in one lot and 23 in the second). These lots were measured before and after stripping (for our discussion, we shall consider only Section B-B, the worst of three sections) and the frequency distributions were plotted in Figure 2.

Similar frequency distributions were plotted for Sections A-A convex and A-A concave. Examination of the section plots showed that within-piece sections were ranked as follows:

1. Analysis of the distributions showed within-piece variation as indicated above.
2. Piece-to-piece variation process too wide s = 0.004 in. for B-B convex.
3. Lot-to-lot variation of 0.007 in. between the two lots.
4. Both lots were on the thin side to start with.

Thus, it was discovered: (a) what section of blade was the worst, (b) that large lot-to-lot variations existed, (c) that piece-to-piece variation is too large for tolerance of 0.030-0.050, (d) that the process is centered too low = 0.0075 in. below nominal, and (e) that even if the averages were raised to nominal, the piece-to-piece and lot-to-lot variations would still cause scrap.

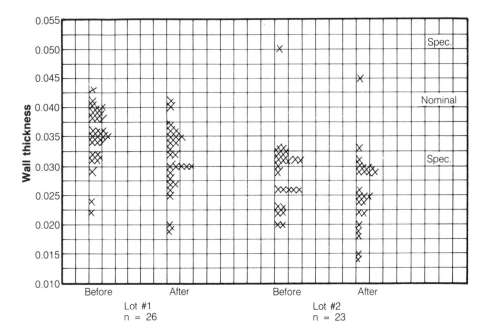

Figure 2 Distribution of wall thickness before and after stripping—Section B-B convex. Stripping analysis for stock removal. Average removal[5] .004 in.

Corrective Action

It was decided to examine blades as received and follow them through the total process, and to perform a distribution study of the process. At this point it was learned that the blades were investment cast in six company-owned molds. It was decided to follow 40 samples per mold from each of the six molds.

Step 2

Samples of 40 blades/mold were measured on all sections and were plotted by mold (Figure 3). A distribution plot by mold (Section B-B convex) shows clearly that molds three and four average about 0.005 in. below the other molds. Mold-to-mold variation, omitting molds three and four, is about 0.0014 in., which is satisfactory. Correcting for mold-to-mold differences, the average is about 0.005 in. below nominal of 0.040 in. Remembering that 0.004 in. is needed for stripping, insufficient stock has been left for reworking the blades.

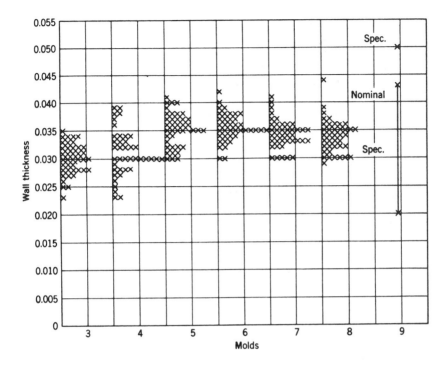

Figure 3 Molds. Analysis of as received material—Section B-B convex.

Conclusions

1. Process averages must be raised.
2. Molds three and four must be separated from others for inspection control purposes.

Corrective Action

1. An engineering change altered specifications on Section B-B to 0.040-.050 to allow sufficient stock for reworking spalled blades.
2. The vendor now submits castings from molds three and four separately for better inspection control. Separation will be continued until molds can be repaired or replaced.

Estimated Savings

Estimated savings are $60,000 a year.

DATA DISTRIBUTION

The value of data collection and its analyses will be realized only if it can be acted upon. This is largely a distribution problem and a matter of getting the information to the right people.

Several techniques may be used to facilitate data distribution. One of the chief methods is to establish a standard distribution for certain types of information. This can be achieved by reporting test data and analyses on standard and multicopy forms which are preassigned to specific individuals or functions. Distribution depends on the type of information being reported and the nature of the resultant action to be taken. Generally the following distribution would be required: originator's copy, cognizant engineer, action department, and a central record copy.

Identification of the proper action department is of extreme importance. Manufacturing/process problems should be directed to manufacturing, and measurement problems to the test or quality operation. If the problem existed in a vendor's facility, the purchasing function would be primarily involved. The Engineering Department is involved in a majority of the problems, regardless of the action department, because of the implications of corrective actions on product performance.

CORRECTIVE ACTION

A company's main purpose of data gathering and analysis is to facilitate decision making. Decisions may vary from not taking any action at all to stopping or discontinuing production. This phase of the cycle is frequently typified by a closed-loop reporting and corrective action system. In companies dealing with the government, you would expect to see a formal system governed by written policy and procedures. Elements of such a system are presented to give you insight into the details involved.

Objectives of the failure reporting and corrective action system are as follows:

1. To promptly document and rectify all failures and malfunctions.
2. To build up and maintain histories of success and failure information to be used for reliability applications.
3. To prevent or minimize the recurrence of failures and malfunctions in design, development, test, and manufacturing phases.
4. To provide management with a visibility of performance of components, systems, and items of equipment (through the media of status reports).
5. To conform to customer and in-house requirements and commitments.

The flow of failure information and corrective action processing through the system is shown in Figure 4.

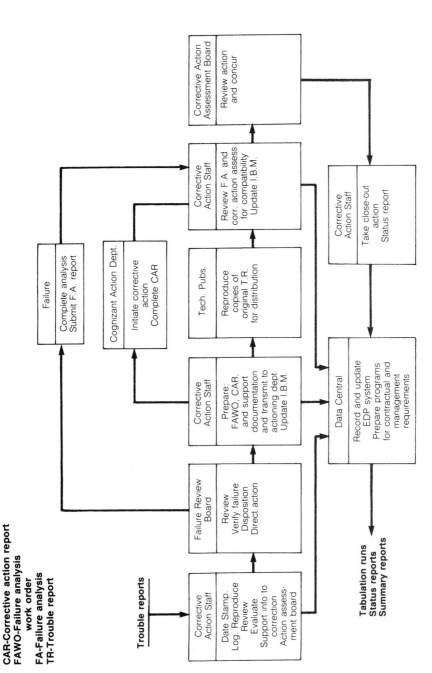

CAR-Corrective action report
FAWO-Failure analysis
work order
FA-Failure analysis
TR-Trouble report

Figure 4 Flow chart of failure information and corrective action system.

All failures, malfunctions, and other troubles are reported by all activities on the specified Trouble Report form. This form is prenumbered to prevent the possibility of assigning the same number to more than one failure. All forms are stored for control purposes by the corrective action staff function.

This system is designed to provide continuous coverage throughout the life of a program through close coordination and interface with design departments and with in-house manufacturing divisions.

Such a system provides the corporation with a library of rapidly retrievable data and support information pertaining to all discrepancies and resultant corrective action. This information is invaluable for building reliability into future products.

The processing and monitoring functions of corrective action are explained by depicting the method of handling a Trouble Report from the time it is received to the time it is closed out.

When a Trouble Report is received, it is date-stamped and logged. Copies are made for review by the Failure Review Board (FRB); the original report goes to the action area of the Corrective Action Staff (CAS) whose responsibility is to evaluate the report for accuracy and completeness.

Initial evaluation includes the following steps:

1. Searching the data bank for similar problems that may provide additional support data to actioning departments.
2. Assigning criticalities (that is, major, minor) to the reported trouble.
3. Coding the report for transmittal to the computer system.

At the conclusion of these steps, a report is submitted to the FRB with recommendations for corrective action.

After a review by the FRB, the report is returned to the action area together with the board's disposition and direction. At this time, the CAS issues (if required) a Failure Analysis Work Order and/or Corrective Action Request to the cognizant actioning department, or closes out the problem if no action is required.

The progress of action is monitored by CAS from the time of initiation to completion by telecon or personal contact with design, quality, test engineers, etc. Corrective action may result in drawing and specification changes. These documents are to review the accuracy and compatibility of the prescribed action. In addition to monitoring and reporting progress of corrective action, the CAS supplies and continually updates information for use in future research.

CAS provides failure data to manufacturing divisions and to design engineering departments, for assessment of potential trouble areas in future products. This information is used to evaluate both systems and component design for existing and new products.

The objectives of an effective Corrective Action System are not confined to correcting an immediate problem; they also ensure that all steps are taken to prevent recurrence. It is essential, therefore, that the corrective action specified be clearly defined and understood to ensure that all related documentation is changed or updated at the earliest possible time.

Many failures result from misinterpretation of specifications and test procedures; subsequent examination of the document concerned indicates that, although the basic intent of the specification or test procedure could be determined, the full definition was obscure because of (in many cases) the incorrect use of wording or lack of detail. Unless these problems are rapidly resolved and complete clarification is given of the intent of the document involved, further failures may also occur.

Failures are also discovered in deliverable (and, sometimes, delivered) hardware. The failure may be such that many identical items from the same series or batch may be affected. As soon as the failure is defined, immediate steps must be taken to ensure that the balance of the items are checked for the known failure mode.

In such cases, CAS has the responsibility to take steps immediately to ensure that the responsible organization is notified of the hazardous condition.

In the early stages of new programs, the CAS reviews the covering documentation for hardware requirements, together with related formal test programs. When the hardware definition is known, the data bank is researched and all failure data (in the form of a computer printout) is submitted to cognizant design areas for review. Thus, problem areas may be highlighted to the designer at a very early stage, enabling an evaluation of the application of parts and components. Although a certain number of failures will occur in any program, every effort must be made to prevent their recurrence.

Success of a Corrective Action System is measured by its effectiveness. Positive identification and definition of a failure are essential to enable the processes of analysis and rectification to be applied. The FRB is established to provide the necessary technical coordination for problem evaluation, and to direct any action required to rectify a problem and prevent its recurrence. Administration of the board is by the CAS. CAS also provides the secretary who is responsible for preparing minutes and presenting Trouble Reports and Environmental Laboratory Failure Reports.

Permanent members of this board are representatives from corrective action and data central, product test and evaluation, and the cognizant product manager. Additional representation is requested from human engineering, product assurance, and design when problems concerning their specific sphere of interest are under review.

Each problem is reviewed, together with any relevant support documentation, and direction is given to the department responsible for failure analysis if required, and for the initiation and implementation of corrective action. Upon completion of a failure analysis and/or corrective action, the report is submitted to the board for review and approval.

The secretary of the board maintains the record of all reported troubles. The record is updated as action processes are completed, and data are instantaneously retrievable for reference purposes. The data are included in the minutes of the meeting, thus providing detailed visibility and continuity of each prescribed action.

SUMMARY REPORT

A monthly report to management summarizes the following activities:

1. Corrective action activity, identifying:
   Action agency
   Delinquency
   Failure classification
2. Failure analysis activity, including:
   Responsible division
   Failure classification
   Delinquency
3. FRB activity, including:
   Total number of failures reviewed
   Initial failure classification
   Initial disposition

The collection of all test reports and preparation of data packages for evaluation by reliability engineering personnel are the responsibility of the CAS. Constant contact is maintained with the various sources of the required information in order that current information may be made readily available for the various tasks involved.

Below are sources of documentation:

   Test reports
   Failure Analysis Reports
   Corrective action reports
   Test plans
   Detailed specifications
   Engineering changes
   Engineering drawings

From the documentation received, the following data packages may be made up for evaluation: test reports, pertinent failure data, and reliability data release forms.

After evaluation, the results are utilized in the following manner: the information completely identifies the test report, part tested, failure data connected with the test, and the mathematical model numbers affected by the parts tested in the program concerned if a system model for the equipment has been developed. Provision is made for cross-reference between the evaluation and the test report.

All test reports and data packages are recorded and filed in the data bank, thus building up a constantly increasing volume of readily available and retrievable test data.

FAILURE RATE DATA HANDBOOK

A further task of the data collection and evaluation group is the preparation and constant updating of a failure rate data handbook. The purpose of this document is to provide the compilation of data in a complete, comprehensive format.

The use of this handbook is explained below:

1. Turn to the "Index" and locate the item number assigned to the part being searched.
2. Locate the item number on the appropriate data page(s). This data will consist of from two to several pages.

The information currently available on each page is as follows:

1. The reliability of the component according to part type and function for all progress combined. This is recorded on the Reliability Assessment Sheet (Figure 5) which furnishes:
   a. Reliability of the part encompassing all programs combined.
   b. Reliability of the part for each individual program.
   c. Source documentation for each individual program.
2. The reliability of the part according to individual program and part number. This is recorded on the Itemized Reliability Data Sheet (Figure 6), which furnishes:
   a. Itemization of test data points according to program, part number, and type test.
   b. Reliability assessment for each part number.
   c. Provides back-up for Reliability Assessment Sheet.
3. Failure data, recording all environmental test failures written against a given component, how they were evaluated by the reliability engineer, why each failure occurred, and how each problem was solved. This type

By _____ Date _____ Subject _____ Sheet No.._____Of ____
Checked by ____ Date ____ _____Reliability Assessment sheet_____Job No._____

Part Name: _____

Part Number: _____

Program: _____

Specific Data

Source	# Units	Successful Events	Failures Crit	Failures Maj	Reliability
Summary					

Remarks:

Figure 5 Reliability Assessment sheet.

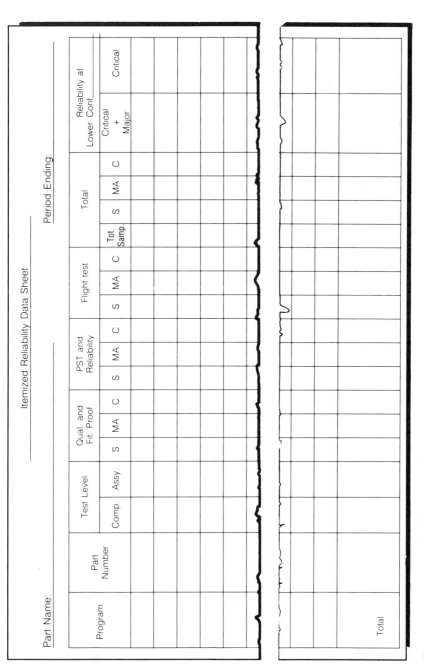

Figure 6 Itemized Reliability Data sheet.

of data is recorded on the Trouble Report Summary (Figure 7), which
furnishes the following on page 133:
a. Reliability data release number (evaluation documentation)
b. Serial number
c. Trouble Report number
d. Failure description
e. Failure environment
f. Reason for failure
g. Corrective action and scoring justification
h. Reliability evaluation [critical failure (CR), major failure (MA), success (S), excluded (E)]
4. Historical data pertaining to design description and function, vendors,
   engineering changes, and detailed specification numbers. Source documentation for the Failure Rate Data Handbook consists of:
   a. Status report—all programs
   b. Test reports
   c. Detailed specifications
   d. Engineering drawings
   e. Engineering changes
   f. Trouble Reports
   g. Failure Analysis Reports
   h. Corrective Action Requests
   i. Reliability data releases
   j. Reliability scoring summaries

   The information derived from Trouble Reports and the action resulting
from them is recorded and maintained on computer files. These files are
constantly updated to reflect the current status until they are closed out with
information displaying the corrective action taken. Through the use of digital computers, the data can be sorted, counted, manipulated, extracted, and
displayed in various formats.
   Transmittal sheets, containing coded and condensed failure data, are submitted
to initiate entries in the data processing system. A detailed listing of the information contained in the data base is presented in Figure 8.
   The data are verified for accuracy and are placed in master files maintained
for each individual product. The files are constantly updated through the
same media of transmittal as previously described. A complete and current
file of failure information is thereby achieved.
   Summaries of the information are prepared periodically and whenever requests
for special information are received. Expansion of condensed information
and interpretation of codes is achieved through the use of digital computers
and computer programs. The programs are designed and constructed to present
the data in a meaningful, easily read format.

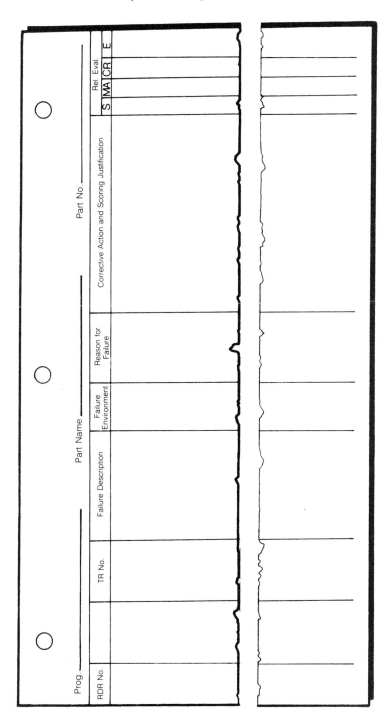

Figure 7 Trouble Report Summary.

CORRECTIVE ACTION DOCUMENTATION & STATUS

Card 1

Program
Trouble Report Number
Initial Trouble Report Number
Reporting Activity
Failed Item Part Number
Part Number Code
Failed Item Serial Number
Failed Item Number
Failure Description Code
Manufacturers Code
Date Failed
Discovered or Occurred During
Reason for Report
Criticality
Degree of Failure
Date Report Received
Report Category
Reference Designator

Card 2

Program
Trouble Report Number
Equipment, Type, Model,
    Designation Number
Equipment Serial Number
Next Higher Assembly Part No.

Figure 8

Next Higher Assembly Serial No.
Problem Identification
Contract
Disposition
Specification
Environment Failed
Test Work Order Number
Small Parts Count
Reference Designator
Card Number

Card 3

Program
Trouble Report Number
Failure Analysis Number
Date Analysis Initiated
Date Analysis Completed
Corrective Action Request No.
Date Corrective Action Request Initiated
Area Responsible for Corrective Action
Date Corrective Action Request Completed
Current Description of Failure
Current Action
Documentation of Action
Effectivity
Date Report Closed
Reference Designator
Card Number

Each week a standard summary of all open problems is issued, which displays the current status of all open items and the areas responsible for the correction action. Each month a summary of items received or actioned in the past six months is prepared.

The use of open-ended coding structures provides a wide flexible base for continued adaptation and adjustment to the specific needs of present and future programs. A central data bank of information, compiled from past programs, provides a media for rapid research.

The tasks of failure evaluation, corrective action, and the preparation of status and summary reports depend on the availability of relevant information required for their conclusion.

The data bank, which is maintained and constantly updated, has been designed to be compatible with the many varied data requirements of both contractual and management reports, and is divided into the following two information storage areas: (1) printed documentation, and (2) computerized data base.

Retrieval and cross-referencing of information between the two systems are easily and rapidly achieved. The major objective has been to store all data, essential to the resolution of problems and for use in the various areas of research and evaluation, for future programs. The printed documentation that is held on file in the data bank is as follows:

Specifications
Specification change notices
Test procedures
Test reports
Quality assurance test procedures
Flight proof certificates
Standards
Schematic drawings
CCB directives
Failure Analysis Reports
Engineering change proposals
Engineering change releases
Engineering changes—Classes I and II
Trouble Report packages (all programs), comprising:
    Trouble Report
    Failure Analysis Reports
    Corrective Action Request
    Action document (for example, EC)
    Specifications change, etc.

All information and data pertaining to Trouble Report packages are retained and updated as required. This information is the source of the data required for research activities in corrective action. The total data bank covers a very broad field of technical documentation that is used in all spheres of product evaluation and corrective action.

REPORTING

An essential adjunct to a failure reporting and corrective action system is effective status reporting that must convey a factual display of a situation for management visibility.

A data storage system, from which status reports are derived, must lend itself to a broad flexibility in retrieval and presentation. Data retrieved must contain sufficient information to provide a clear picture at any given time. The data processing system used should be designed and developed with accent on flexibility in retrieval.

Status reports are produced in a variety of formats by means of the extensive coding system employed. Coding lists are shown in Figure 9. Additionally, tabulation runs can be supplied (on very short notice) on specific items, part numbers, failure modes, environments, and the like.

The following summaries are produced and distributed on a routine basis:

1. Open Trouble Reports—weekly.
2. Trouble Reports and Corrective Action Requests—monthly.
3. Reliability Status Reports—monthly.
4. Status of Corrective Action Assessment Board (CAAB) actioned items— monthly.
5. Status Summary of Environmental Tests—weekly.
6. Summary of Discrepant Material Reports (DMRs)—monthly.

The above reports are issued on preprogrammed standard formats. Reporting of tests represents a different problem from that of reporting failure information. Development of format and content criteria to be used in preparation and careful evaluation. General requirements encompass all collected data and all resultant conclusions. An outline of a complete test report includes the following:

a. Title page and cover. Identify the report by name and number. Indicate the item tested by type, size, rating, etc., and the testing agency.
b. Table of Contents.
c. Reason for test.
d. Description of test specimen. Describe tests on each type of component. If possible, restrict test coverage and report to only one type. Identify the item tested in detail, i.e. serial number, part number, rating, and control numbers. Also, state basis for sample selection, state of item development, and brief functional description.
e. Disposition of test specimen.
f. Conclusions and recommendations. Briefly summarize the results of this test; state any corrective measures taken; suggest improvements for further consideration; indicate all corrective actions to be taken in the event of failure.
g. References.
h. Failure-retest provisions. Specify the actions to be taken in event of failure.

FAILURE CODE—ALPHA

A

001 Activates Incorrectly
146 Adjustment, Out of
139 Age,—Old
088 Alignment—Improper
002 Arcing

B

004 Bent
005 Binding
006 Blistered
007 Blown
008 Bonded
106 Break in Insulation
084 Breakdown in High Voltage
010 Broken
169 Broken Seal
186 Build-up of Tolerance
012 Burned
053 Burred

C

013 Calibration Incorrect
014 Capacitance Incorrect
015 Chafed
016 Changed Value
097 Charged—Insufficiently
017 Charred
018 Chatter
197 Chipped
019 Clogged
022 Closed,—Fails to
020 Cold Solder Joint
021 Collapsed
089 Connection—Improper
046 Consumption
157 Contact—Poor

023 Contacts—Defective
027 Contamination
025 Corroded
026 Cracked
027 Current—Incorrect
198 Cut
056 Cycle,—Fails to
090 Cycling Improper

D

028 Damaged
029 Defective
023 Defective Contacts
030 Defects,—Manufacturing
168 Defects,—Screen
192 Delaminated
031 Dented
205 Design Problem
032 Destroyed
034 Deteriorated
052 Diagnosis Incorrect
147 Dimension,—Out of
024 Dirty
036 Distorted
200 Drawing Incorrect
037 Drifts

E

202 EC Not Incorporated
115 Elements Loose
049 Energize, Fails to
091 Engagement Improper
099 Equipment Insufficient
040 Erratic
096 Error, Human
043 Excess Gap
045 Excess Load

Figure 9

204 Excessive Use
047 Excessive Vibration
048 Exploded

F

022 Fails to Close
056 Fails to Cycle
049 Fails to Energize
050 Fails to Fire
051 Fails to Initiate
054 Fails to Open
055 Fails to Position
057 Fails to Regulate
060 Fails to Separate
061 Fails to Stop
062 Fails to Switch
063 Fails to Tune
065 Fails to Zero
066 Fatigue
095 Fault Indicated
068 Faulty Inspection
067 Faulty Material
069 Finishing
050 Fire, Fails to
071 Flaking
072 Fluctuating
074 Frayed
075 Frozen
077 Fused

G

119 Gain-Low
043 Gap—Excessive
079 Gouged
080 Grooved
081 Grounding—Improper

H

092 Handling—Improper
085 High VSWR

084 High Voltage Breakdown
086 Human Error
087 Hysteresis

I

093 Identification—Improper
102 Impedance—Incorrect
088 Improper Alignment
089 Improper Connection
090 Improper Cycling
091 Improper Engagement
081 Improper Grounding
093 Improper Identification
094 Improper Installation
122 Improper Lubrication
003 Improper Operation
206 Improper Packaging
009 Improper Usage
011 Improper Wiring
098 Inadequate Test Equipment
001 Incorrect Activation
013 Incorrect Calibration
014 Incorrect Capacitance
027 Incorrect Current
052 Incorrect Diagnosis
200 Incorrect Drawing
102 Incorrect Impedance
149 Incorrect Output
042 Incorrect Pulse
041 Incorrect QATP
164 Incorrect Resistance
196 Incorrect Test Specification
183 Incorrect Timing
184 Incorrect Tolerance
185 Incorrect Torque
190 Incorrect Voltage
095 Indicates Fault
051 Initiate, Fails to
096 Inoperative
068 Inspection Faulty
094 Installation Improper

Figure 9 (*continued*)

099 Insufficient Equipment
100 Insufficient Insulation
097 Insufficient Charge
101 Insufficient Protection
106 Insulation Break
039 Insulation Resistance
107 Interference
108 Intermittent
109 Insulation Resistance

J

112 Jammed

L

165 Leads Reversed
113 Leakage
135 Linear,—None
045 Load Excessive
114 Loose
115 Loose Elements
119 Low Gain
121 Low Sensitivity
122 Lubrication—Improper

M

030 Manufacturing Defects
067 Material Faulty
160 Measure Preventative
120 Melted
124 Mismatched
123 Missing Document
125 Missing Part(s)
126 Moisture

N

195 No Failure
203 No Output
132 No Response
133 No Solder
129 Noisy

135 Non Linear
136 Not Determined

T

182 Technical Procedure
098 Test Equipment Inadequate
196 Test Specification Incorrect
180 Threads—Stripped
183 Timing—Incorrect
186 Tolerance Buildup
184 Tolerance—Incorrect
185 Torque—Incorrect
063 Tune,—Fails to
187 Test Equip. Fail

U

188 Unbalanced
033 Unbonded
203 Underweight
189 Unstable
204 Use, Excessive
009 Used Improperly

V

085 VSWR High
016 Value Changed
047 Vibration, Excessive
158 Voids,—Potting
084 Voltage Breakdown High
190 Voltage Incorrect

W

191 Warped
011 Wiring Improper
193 Worn
194 Wrong Part

Z

065 Zero, Fails to

i. Test data. Include pertinent and complete information for the following:
   Test method description
   Test procedure
   Summation and analysis of test results
   Exact test data in the form of measurement taken
j. Appendixes. Include pertinent information that will aid in effective use
   of the test report.

SUMMARY

We have discussed data requirements, the rationale associated with determining the data required, the form it should take, the technique and analysis, and the method of reporting and distribution. Let us emphasize this point: regardless of the product or type of data available, the key factor is the purpose that it is to serve and the value to be derived from it in making a decision. Proper identification of its desired use will dictate the various aspects of the data system to be established. Elements in the procedures described must be geared to the size of the company and to the program or product to which it is to be applied.

RECOMMENDED READINGS

Lloyd, David K. and Myron Lipow. *Reliability: Management, Methods and Mathematics.* 2nd ed. Redondo Beach, Calif., 1977.

Lucas, Henry, Jr. *Information Systems Concepts for Management.* 2nd ed. New York: McGraw-Hill, 1982.

Ein-Dos, Phillip and Eli Segev. *A Paradigm for Management Information Systems.* New York: Praeger Publishing, 1981.

Murdick, Robert G. *MIS, Concepts and Design.* Englewood Cliffs: Prentice-Hall, 1980.

Chapter 6

Formulating and Implementing the Integrated Evaluation Program

In a test effort, it is essential for management control purposes that a systematic integrated evaluation program be formulated and implemented to supply the information needed by the PT&E data system. Although every product has its own particular requirements, specific guidelines that offer assistance in making applicable decisions must be established. The material in this book points out the importance of test and evaluation from the inception of product design to delivery of the final product and, in many instances, follow-up testing when service life and reliability are factors.

The integrated test program concept is the most comprehensive method of mating the planning and the required tests, and correlating the resulting test data. A properly integrated test program defines the constraints, objectives, responsibilities, and techniques for implementing, monitoring, and controlling the test and evaluation of the product, and provides a system for disseminating and utilizing data from all test sources. The program also reflects observance of the policies and capabilities of the company (Manufacturer & Producer of Product).

An integrated test tree, defining such a program (collating the PT&E management information requirements of Chapter 5), is shown in Figure 1. It is a most useful tool in depicting the complete scope of the test program.

OBJECTIVES AND REQUIREMENTS OF THE INTEGRATED TEST PLAN

To ensure timely, efficient, and economical implementation of the test portion of the integrated evaluation program, careful coordination is required.

An integrated test plan, which serves as a road map for all personnel and which defines and controls the relationship of all testing, must also satisfy the following objectives:

a. Detect engineering deficiencies as early as possible to permit suitable corrective action.
b. Test in the most realistic operation environment available.
c. Eliminate redundancy of test effort.
d. Provide as much flexibility as possible to allow for inevitable changes which cannot be predicted, for example, failure, with subsequent retest requirements.
e. Ensure that critical test items will not be subjected to excessive or unnecessary operation.
f. Promote orderly progression from laboratory test through operational use on schedule.
g. Optimization of test specimen cost through maximum use of all test specimens.
h. Continuous control of test specimen configuration.
i. Provide a basis for the correlation and utilization of test data from all sources.
j. Permit orderly and controlled implementation of the test program with logical timing, sequencing, and interrelation of the several types of tests.
k. Correlation of discrepancies and failure experience from test to test, but primarily from development through final acceptance testing.
l. Orderly development and verification of requirements and procedures in parallel with the hardware development.

Test Integration Plan Contents

To fulfill the objectives and requirements of the integrated test program, a formal control plan must be written. This plan establishes what must be done, who must do it, and when. It describes in detail the product to be tested, the environments that it will be subjected to, and the test schedule. The PT&E organization is responsible for writing the plan, overseeing test implementation, and pinpointing personnel responsibilities.

Integrated test plan formats will vary, depending on the individual company structure and the item being tested. However, for successful integration, the basic integrated test tree of Figure 1 should be used. It also serves as a guide for writing the necessary documents. A more comprehensive plan can be written by first carefully outlining the diagram in Figure 1 and then describing the purpose and requirements of each block, as will be explained in this chapter.

Detail Test Plans and Procedures

Prior to delivery of test specimens, a detail test plan must be prepared for each item to be tested. With the aid of the detail specification for the item

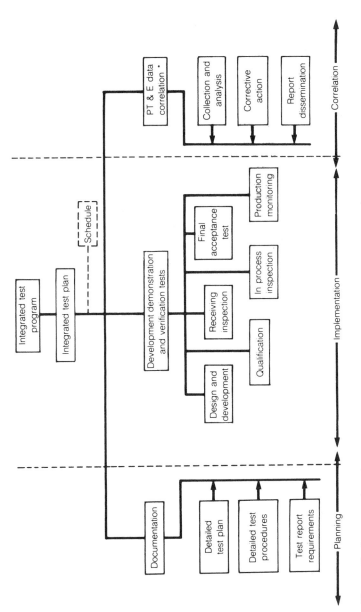

Figure 1 Integrated testing tree.

and the test program plan, and depending on the type of test, sufficient information should exist to outline a comprehensive plan. Some elements of the detail test plan are given below:

1. Introduction. States the purpose of the test, cites applicable reference documents, identifies by part number and describes the intended function and operation of the item as well as the number of test specimens.
2. Requirements. Cites the applicable specification, performance, functional, and environmental test requirements.
3. Test Procedure. Lists and describes in detail *how* the inspection and bench test measurements will be made, specifying instruments, fixtures, circuit diagrams, and performance measurements methods, designating when (before, during, or after environmental exposure) these measurements will be made. This section describes the method and sequence of the individual environmental tests, test setups, and test facilities. Also described, after environmental tests are completed, are the specific methods for evaluating the effects of the environmental exposures on the item, including autopsy, X-ray, and other nondestructive tests.
4. Evaluation. Describes the criteria for test success which may be on a go-no-go basis, or may require demonstration of a confidence level from statistical analysis of the data.
5. Appendixes. May be provided to illustrate sample data sheets, details of fixtures, setups, circuit diagrams, and any other relevant material.

Test Report Requirements

Test reports are prepared for each test or group of related tests in each test category. These reports are based on the results of all analyses performed on data obtained during the testing, and emphasize the results of testing relative to each test objective; they indicate whether compliance with design of specification requirements has been achieved, and provide recommended changes to equipment to affect required design and reliability improvement.

All pertinent data, graphs, charts, photographs, and references (including inputs from failure, and statistical analysis support activities) are included. Thus, total technical justification is presented to support the analysis and recommendations in order to utilize the test results to improve component and system reliability.

DEVELOPMENT, DEMONSTRATION, AND VERIFICATION TESTS

Correlating the actual testing into the overall integrated test program must be done judiciously in coordination with the organizations that will be cognizant for test performance. To implement the program plan the following types of testing are recommended: (a) design and development testing to select materials and parts to be used and to improve designs, (b) qualification testing to demonstrate design capability and reliability of prototypes, (c) receiving inspection to assure quality of purchased or subcontract items, (d) in-process acceptance testing to assure satisfactory workmanship, (e) final operational production testing to maintain design capability and reliability and to assure satisfactory workmanship, and (f) production monitoring to check workmanship and performance characteristics.

Design and Development Testing (DDT)

The initial phase of parts testing takes place in the design and development phase and is carried out as part of the initial parts engineering activity. Work during this phase includes background studies with parts characteristics, comparative checks to aid designers in making initial choices of parts for prototype assemblies, and studies of specific problems that may appear later in a program.

The amount of design and development test evaluation performed in support of a given product development program depends on the background and facilities of the firm and on the availability of applicable test results.

If extensive information on the prior use of a part under similar operating conditions is available, many of the answers that the design and development program would provide can be anticipated. In this case the testing would be limited to a few specialized problems. If such information is not available, designers may test to obtain answers not readily available in their experience or elsewhere, and a larger number of the part types to be used must be tested. This suggests the merits of extensive exploration of the data sources previously mentioned in order to minimize the need for such effort.

Sample sizes are determined by requirements for locating and studying major parts defects such as susceptibility to specific environmental conditions associated with operational use and storage. Sample sizes on the order of 30 or each part type with relatively high reliability requirements, or ten of each part type with low reliability requirements can be considered adequate. The discussions on statistical considerations and close coordination with the PT&E engineering section may further influence selection of the final sample size.

Qualification Testing

After initial choices of parts are made, comprehensive qualification tests are carried out as a basis for subsequent procurement and use. These tests are carried out according to stringent procedures prepared specifically to ensure rigid uniform evaluation results. Qualification test results, are used here, include: (1) verification of performance capability of a particular design for the intended application, (2) the use of environments to determine that no serious weaknesses exist that could cause inconsistent performance or high probability of failure, and (3) a basis for evaluation of assembly and component test results. Initial parts qualification should be completed in time for use of qualified parts in the final qualification test of the system, if a system is involved. Since qualification of the part at a higher level is a function of the design and production processes, periodic requalification may be required because of production and/or design changes.

The amount of qualification testing (or testing of this nature) to be performed in support of product development may also be influenced by supplementary information. If considerable experience with a part has led to qualification for another application and the proposed application is similar, the part may be utilized without duplicating the test effort. Again, in this case, documentary evidence from previous test data may be utilized to support the decision (choice of part). Valuable sources of data are vendors if they have conducted and documented comprehensive parts testing programs, particularly when such programs have been carried out by independent test laboratories. Utilization of results such as these will accelerate parts improvement efforts, and will tighten overall economy by increasing general applicability and availability of test results.

Receiving Inspection

Acceptance tests of parts are carried out to check the quality of purchased incoming products. Test objectives are to determine that lot product quality is satisfactory and that proper production processes are being maintained. Because of large sample size requirements and the time involved, it is usually impractical to check reliability by testing. Direct in-plant monitoring and review of in-process test and production process data are usually applied to determine conformance to these requirements. Sample sizes appropriate to the situation normally can be found by utilizing various sampling plans available.

In-Process Inspection

Testing of assemblies, as with parts testing, is carried out to support initial designs or redesign. Assemblies include all aggregations of piece parts that are not necessarily functional entities in and of themselves. This distinction differentiates between this and equipment testings discussed under the heading Final Acceptance Test.

Objectives of assembly testing included background studies of material application, experimental models, or fabrication techniques, initial checks of completed assembly designs, tests of piece parts more advantageously tested in assemblies than tested separately, and investigations of special problems that may arise during the program.

Sample sizes for assembly testing are determined by the problems and characteristics to be explored. Sample sizes on the order of three assemblies may be considered adequate when the problems are not too significant and when equipment testing is also planned. However, if an assembly is critical because of special problems, more exhaustive testing should be done to preclude future trouble at the equipment level.

Final Acceptance Test

Equipment testing (complete functional items) is of particular concern because (1) it represents the highest assembly level of test activity, (2) it involves a maximum of subsystem and environmental interactions for development study and evaluation, (3) environmental test conditions applied to equipment are most readily related to measurements of end used conditions, and (4) equipment testing requires a maximum of advance planning for production, repair, and test facilities.

Equipment testing usually represents the final stage of laboratory debugging and evaluation. Such testing normally progresses through the following two-phased groups: phase (a) identification of major modes of failure and degradations in performance caused by environment, engineering analysis of failure modes and performance degradations, and correction of design as appropriate; phase (b) determination of the importance of remaining failure modes and performance degradations, investigation of effects of wearout on reliability and performance, analysis of reliability and performance extrapolation of results for estimation of reliability under service conditions, further correction of design where necessary, institution of needed preventive measures, and correlation with acceptance tests.

The amount of equipment testing planned should be sufficient to allow location of significant modes of failure. The program should provide for the investigation of effects of varied shock levels, combinations, and sequences of environments. Again, the test program planners should apply the techniques of statistical design of experiments to capitalize upon the testing to be performed.

The number of items to be tested should be sufficient to allow thorough debugging and to provide moderate statistical confidence in the results of the test program. Such confidence should be established on the basis of repetitive testing where the product permits. The reliability characteristics of the product may prove too demanding from a financial and time point of view to attempt to verify during this phase. It is suggested that from five to ten equipments of each distinct design be allocated for this test category.

Final acceptance tests normally are performed at the factory, when the equipment is procured, on each developmental equipment before delivery as a means of improving reliability by revealing workmanship defects and incipient flows for correction prior to end use. The tests are not intended for design evaluation. Test levels applied should be comparable to those expected in operational use and should be well within design limits. The tests should be long enough to produce initial failures but not long enough to induce equipment wearout. These tests should advance the equipment beyond the period of high initial failure rate of its life failure curve to the beginning of the middle life period of maximum reliability.

Production Monitoring

Once the product is authorized for production, the workmanship and performance characteristics must be checked. This is the purpose of the production test program. A combination of nondestructive and destructive testing may be employed. Of course, the destructive testing must be performed on a sampling basis. The sampling plan should be selected on the basis of risk, cost, and reliability requirements. As production experience on the specific equipment is gained and maintenance of quality is demonstrated, test and sampling procedures may be established involving sequential sampling revision of test levels or other variances considered appropriate.

PROGRAM SCHEDULING

Program scheduling for all system testing permits the evaluation of test effectiveness in terms of hardware, facility, and personnel availability. Elements of the schedule are the delivery data for the test specimen, release date of the detail test plan, starting and completion dates for testing, and release date of the test report. The delivery date of the test specimen is contingent upon the design release date and required procurement lead time. Detail test plans must be released before testing starts. Test starting dates are planned immediately upon receipt of test specimens but may be governed by manpower and test facility loadings. The total time required to complete testing is estimated by the test engineer and is governed by the type of test, the number of specimens to be tested, complexity and number of functional and passive performance measurements, and allowance for delays due to minor failures of the item and facility usage conflicts. An allowance for preparation of the test report should be included in the schedule.

An overall completion date is thus etablished for an entire test program based upon the detail completion dates. Scheduling may be accomplished by rearranging the blocks of Figure 1 and adding the "need dates" (time requirements) or by a simple PERT (program evaluation review technique) system.

SUMMARY

We have outlined an integrated system approach to product test and evaluation which, when applied, will result in a most comprehensive program of system development and operational verification. Of particular importance is the systematic approach to definition and the correlation of the detailed plans and tests of the integrated program (which may be adapted into a program schedule). Also stressed are the considerations required relative to test facilities and the applicability of prior experience to permit maximum use of the concept "qualification by similarity," which is vitally important in an accelerated program. An integrated system approach to product evaluation is a concept that can contribute significantly to product knowledge and, at the same time, can produce substantial savings in total evaluation costs and time.

RECOMMENDED READINGS

Grant, Ireson W. ed. *Reliability Handbook.* Sect. 8. New York: McGraw-Hill, 1966.
IEEE Transactions on Reliability, August 1983
  Reliability Programs for Commercial Communications
  Satellites, F. E. Erdle et al.
  Reliability-Growth Programmes for Undersea Communication Systems, R. H. Murphy.
  Planning of Production Reliability Stress-Screening Programs, Fiorentina & Saari.
  Managerial Decision-Making in Establishing R&M Design Goals, A. F. Czaikowski.
  Reliability Investment and Life-Cycle Cost, J. K. Seger.

Chapter 7

Environmental Tests:
Temperature, Humidity,
Mechanical Shock, Pressure, and Fungus

ENVIRONMENTAL TEST PHILOSOPHY

The recent breathtaking advances of research and PT&E have resulted in increasingly complex techniques and methods in modern test and evaluation work. Simulated natural and induced environmental tests for evaluating product performance have evolved gradually over a long period of time. Here, we discuss tests that have been developed scientifically and those that have been discovered through trial and error. With the coming of the space age, the simulation of certain environments is a challenge to both earthbound facilities and application engineering. But advances in test methods are not restricted to space programs; evaluation methods of commercial products have also been improved.

The test methods contained in this chapter and Chapters 8 and 9 specify suitable conditions, obtainable in the laboratory, that give test results similar to actual service conditions, and guide those engaged in preparing the environmental test portions of individual equipment specifications.

When selecting tests, we should carefully consider the anticipated environmental conditions. Conditions that reflect actual service usage, including shipping and ground handling, should be taken into account. Conditions that would adversely affect or very probably induce a malfunction of the test item should be given special emphasis.

Operational Checkout (Before Environmental Exposure)

Before any of the tests are carried out, the test item should be operated under standard ambient conditions, and a record should be made of all data necessary to determine compliance with required performance. These data will provide the criteria for checking satisfactory performance of the test item during or at the conclusion of the test.

The test item should be installed in the test facility at room temperature in the axis that will simulate service usage. Plugs, covers, and inspection plates, used in service, should remain in place. When mechanical or electrical connections are not used, those normally protected in service should be adequately covered. The test item should then be operated to determine that no malfunction or damage was caused by faulty installation or handling. The requirements for electrical and/or mechanical operation, following installation of the test item in the test facility, are applicable only when the operation is required during exposure to the specified test.

When electrical/mechanical operation of the test item is required during the test exposure, the operation and performance checks should be of sufficient duration or should be repeated at appropriate times and intervals to insure a comprehensive record of all data for comparison with data previously recorded.

The test item should be visually inspected, and a record should be made of any damage resulting from the test. Normally the test item should be removed from the test facility before inspection, as stated in the various test methods. However, when the installation of the test item in or on the test facility is complex, costly, or time consuming, the test item may be inspected inside or on the test facility, provided that all inspection criteria are met. (If a test chamber is used for the test and the inspection is performed inside it, the chamber is returned to conditions of room ambient temperature, atmospheric pressure, and relative humidity before proceeding with the inspection.)

Deterioration, corrosion, or change in tolerance limits of any internal or external components that could, in any manner, prevent the test item from meeting operational requirements should be considered as proof that the test item cannot withstand test conditions.

Allowable Tolerances

Unless otherwise specified, all measurements and tests are made at room ambient temperature, atmospheric pressure, and relative humidity. Whenever these conditions must be closely controlled in order to obtain reproducible results, a reference temperature of 23°C (73°F), a relative humidity of 50%, and an atmospheric pressure of 30 in. of mercury, respectively, should be used together with whatever tolerances are required to obtain the desired precision of measurement. Actual ambient test conditions should be recorded periodically during the test period.

The maximum allowable tolerances of test conditions (exclusive of accuracy of instruments), except as stated in any one of the test methods of this chapter or as stated in the individual equipment specification, are as follows:

1. *Temperature.* ±2°C (3.6°F).
2. *Pressure.* When measured by devices such as manometers, ±5% or 0.06 in. of mercury, whichever provides the greatest accuracy. When measured by devices such as ion gages, ±10% to 1×10^{-5} torr.
3. *Relative humidity.* Plus 5% R.H., minus 0%.
4. *Vibration amplitude.* Sinusoidal ±10%. Random ±30%.
5. *Acceleration.* +10%.

The accuracy of instruments and test equipment used to control or monitor the test parameters, whether located at an independent testing laboratory or at the producer's plant, should be verified periodically [at least every 12 months, preferably once every six months, unless producer procedures prepared to satisfy the requirements of MIL-STD-45562, MIL-Q-9858 or NASA-NHB-5300.4(1B900) for calibration cycle of specific instruments specify otherwise] to the satisfaction of the procuring activity. All instruments and test equipment[1] used in conducting the tests should:

1. Conform to laboratory standards whose calibration is traceable to the prime standards at U.S. Bureau of Standards.
2. Have an accuracy of at least one-fifth the tolerance for the variable to be measured. In the event of conflict between this accuracy in any one of the test methods of this chapter, the latter should govern.
3. Be appropriate for measuring the environmental conditions concerned.

TEMPERATURE TESTS

General guidance and suggested minimum requirements for conducting temperature tests within the range of −80 to +200°F are given. Temperature chamber testing consists of subjecting one or more test items to various temperatures listed in the detailed test specifications. Five frequently used standard temperature chamber tests are constant high-temperature test, high-temperature cycling tests, constant low-temperature test, low-temperature cycling test, and temperature shock tests.[2]

Most difficulties encountered in conducting a temperature chamber test result from one or more of the following conditions: too many test items in a chamber, thermal lag of chamber walls, wide variations in temperature within the chamber, inaccurate temperature controls, inadequate methods for determining temperature stabilization, frosting of electrical contacts, and improper spacing of test items in the chamber.

The usual practice is to place the test item in a chamber already at or near the test temperature. This procedure applies a thermal shock to the test item. If this procedure would be detrimental, an alternate one should be specified. Normally, "exposure time,"[3] rather then "stabilization time,"[4] is the preferred term and is the test time referred to in detailed test specifications.

Test Specification Check List

Certain information is required to assure a properly conducted temperature test. The following points should be listed or considered in the preparation of the test specifications, and the following items must be specified:

1. *Test temperature(s).*
2. *Test temperature tolerances.* A recommended absolute tolerance is ±10°F within the range of −80 to +200°F. (For temperatures outside this range, an absolute tolerance of ±10°F is difficult or impossible to attain.)
3. *Exposure time.* If critical, specify the minimum and maximum exposure time. A recommended tolerance for time is ±10% of the duration specified.
4. *Type of protection against moisture condensation and frost.* If protection is required, so specify and describe the means in detail. Under certain atmospheric conditions, moisture may condense on electrical connectors after protective coverings have been removed.
5. *Functional tests.* The tests to be performed, as well as the environmental conditions and time intervals, should be described. If the test item can be removed from the chamber to facilitate functional testing, this should be noted, and a maximum allowable time for this operation should be specified.

The following items need not be specified but should be considered:

6. *Relative humidity.* Specify only if it is a factor.
7. *Number of sensors.* Specify where and how many, if required.

8. *Temperature of chamber.* Must the chamber be at room temperature before the test item is introduced? The normal procedure is to introduce the test item into a preconditioned chamber. This procedure applies a thermal shock to the test item. If this procedure would be detrimental, an alternate one should be specified. Deviation from normal practice is costly and time consuming and should be fully justified.

Instrumentation

The control instrumentation of a chamber should be accurate, sensitive, and quick in response to temperature changes. A chamber should have at least one accurate indication system. Recording may be either continuous or periodic as specified. The temperature indication should represent the condition at the reference point. The temperature distribution measurement should indicate whether corrective action will be necessary. Frequent spot checks of the temperature indication system can prove its stability or detect trouble before a test is invalidated because of improper temperature control. Spot checks should be performed with an independent measuring device.

Accuracy of calibration depends on an accurate transfer standard, competent personnel, proper location of temperature sensor, and care in the handling and use of the transfer standard. The personnel assigned to perform calibration should be both competent in the use of calibration instruments and interested in accuracy and precision, and then must know how to attain the degree of precision required. The temperature sensor of the chamber control instrumentation may be accurate and capable of controlling the temperature at its location, but this temperature may not be the same as that of the rest of the chamber volume. Thus the location of the standard sensor is important. From the initial temperature distribution measurements, a point representative of the chamber temperature may be selected for the location of the standard sensor.

The frequency of calibration for a standard depends on its stability. Many devices available today are inherently stable for considerable periods if they are carefully handled. Thus care in the use of such devices is important. It is recommended that when a device (glass-stem thermometer, thermocouple/potentiometer, resistance thermometer) is designated as a standard, its use be restricted to calibration functions. Furthermore, only specified persons who will take meticulous care in their work should handle the device.

Recommended standards in the order of accuracy preference are: (1) the resistance (platinum) thermometer, (2) the liquid-filled glass ther-

mometer (with certification by an authorized standards laboratory), and
(3) the thermocouple (Cu-const. or Plat. −10% Plat. Rhod.) and poten-
tiometer. The resistance thermometer and the thermocouple, although
more expensive than the glass thermometer, are more versatile.

Test Operation and Equipment

1. *Chambers.* One requirement for a temperature chamber is that
it have the capability of changing and maintaining the temperature
within the tolerances specified. It must have enough reserve power to
perform satisfactorily when subjected to sudden changes in load (as in
temperature shock tests) or in temperature requirements (as in high- or
low-temperature cycling tests). In a well-conducted temperature test,
heat transfer is accomplished primarily by convection, but radiation
from chamber walls may not be negligible. Therefore, temperature distribution within the chamber is seldom, if ever, uniform. To minimize
the deleterious effects of this condition on a test, the ratio of the chamber volume to the total volume of the test item(s) should be as large as
practicable, and if more than one test item is to be placed in the chamber, they should be spaced far enough apart to permit adequate circulation of air.

Although radiation between the chamber walls and the test item(s)
may be negligible, care should be taken to place the test item(s) as far
as practicable from a chamber wall. In other words, the test item(s)
should occupy the central volume of the chamber.

For a chamber operated continuously at one temperature, the test
item(s) may be placed relatively close to, but not in contact with, a
chamber wall. The effect of chamber wall temperature lag (or inertia) is
probably greatest during the temperature cycling tests, when the walls
may act as radiators or heat sinks. If experiments indicate that the
walls will have a measurable effect on the conditions of a test, radiation
shields or baffles may have to be installed in the chamber.

Forced air circulation is necessary in a temperature chamber to provide a better distribution of conditioned air and thus reduce the temperature gradients. In addition, temperature fluctuations should be reduced
to insure that each test item will be exposed, within the allowable tolerances, to the specified temperatures. Temperatures in the test areas of
the chamber should be periodically checked.

2. *Temperature Control and Indication.* For close control, there
are two requirements: the instruments must be accurate and must
respond readily. Accuracy is assured by periodic calibration, which
must be correctly and frequently performed. It is a good policy to
spot-check the instrumentation between calibration intervals to insure
stability and reliability. If thermocouples are used for temperature con-

trol, premium-grade copper-constantan thermocouples are recommended. These are available with an accuracy of ±0.75°F or better.

The control system must respond quickly to maintain the temperature within the test tolerances. Lags within the tripping (or switching) mechanism or within the sensing system can occur. In addition, the location of the sensor in the chamber may influence the response.

3. *Temperature Distribution Measurement.* All areas of the temperature chamber may not be at the specified temperature; thus we should measure temperature distribution at stabilized temperatures. The recommended method does not measure the temperatures directly but, instead, measures the differences between a reference point and other points. A thermocouple is arranged in such a manner that one junction is located at a point of reference and the second junction is located at a the point to be measured (see Figure 1). The advantages of this method are: (1) the readings are plus or minus in relation to a reference point, (2) many possible errors or corrections are either minimized or eliminated, and (3) the recording of data is simple.

Figure 2 is a suggested scheme for multiple-point measurements utilizing "delta" thermocouples. The common point of the constantan leads should be in good electrical contact with each lead. This common point may be located at any convenient place, either inside or outside the chamber. Any number of points may be measured by simply adding or subtracting thermocouples. Care should be taken to make sure that no leads are electrically shorted. Since the thermocouple leads to the switch are copper, the parts of the selector switch should be made of copper or all terminals on the selector switch should be at the same temperature. The temperature of the two connections to the measuring instrument should be equal.

The readout system should be capable of measuring accurately (±10μv) the small potentials generated by the thermocouples.

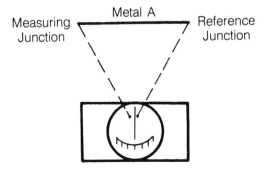

Figure 1 Measuring instrument, delta thermocouple.

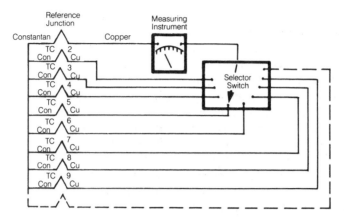

Figure 2 Suggested schematic for multiple-point delta thermocouple array for any number of thermocouples.

The reference junction should be located at the reference point, the geometric center of the chamber usable volume. The temperature difference between the reference point and any distribution check point in the usable volume should be recorded to within ±0.5°F or better. A minimum of eight points will be measured with respect to the reference point. These are the eight corners of a square prism encompassing the usable volume. In addition, a sufficient number of points may be measured to provide an adequate "picture" of the temperature distribution. These additional points should be located at midpoints of the edges of the prism and at the center of the six prism surfaces. Depending on the size of the chamber, other points located in a regular pattern within the usable volume may be measured as necessary. The points must be sufficiently far from the walls to reduce radiation effects but close enough to adequately picture the distribution(see Figure 3).

A rack to hold the array of the thermocouples should be lightweight, have a minimum effect on airflow patterns, and be made of a material with a low coefficient of heat conductivity. Wooden dowels not only meet these requirements but are readily available and economical. Such a rack is convenient if many chambers, having similar interior dimensions, are to be checked. It will also afford some accuracy and repeatability in the placement of thermocouples for successive temperature distribution measurements. Figure 3 is a sketch of a rack.

Before the thermocouples are used, they should be calibrated. First, both junctions of a pair are exposed to the same temperature environment. The indication should be zero. Second, one junction is exposed to a temperature different from the one to which the other junction is being exposed. The difference should be from 3 to 10°F and must be

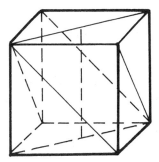

Figure 3 Sketch of rack configuration suitable for thermocouple mounting. (Dotted lines used for clarification.)

accurately measured. This calibration should be performed at the extremes of temperature in which the thermocouples will be used.

A simple check ensures that the system (thermocouple array and readout) is operational. With all thermocouples junctions at the same (room) temperature, each measuring junction is grasped, one at a time, between the thumb and forefinger. The small temperature difference thus created between a measuring junction and the reference junction will cause the readout device to indicate a positive temperature differential if the system is correctly assembled and operational.

Readings should be taken after the chamber is stabilized at the desired temperature. The actual temperature of the central reference point should be recorded at the start and end of each series.

4. *Methods for Determining Temperature Stabilizing.* There are several techniques for determining stabilization time. Some give erroneous or inconsistent results, but the following methods are reliable:

a. The preferred method is to place a thermocouple inside the test item at a point judged to have the greatest thermal inertia. The item is then placed in a chamber preconditioned to the specified test temperature. The time it takes for the temperature, as recorded by the thermocouple to reach within 2°F of the chamber test temperature is defined as the stabilization time.

b. When the preferred method is impracticable, stabilization time may be calculated from qualified analytical or computer heat-transfer studies.

c. When methods (a) or (b) are impracticable to apply, the stabilization time should be determined as follows:

Several thermocouples, shielded from direct air circulation as well as from radiating surfaces of the test chamber, should be installed in good thermal contact with the surface of the test item. The time required for the surface temperature, as recorded by the last

thermocouple, to reach within 2°F of the chamber test temperature, multiplied by three, is defined as the stabilization time. A
prototype model may be used for this determination.

5. *Shielding of Electrical Contacts and Various Test Items.* All test
items and temperature-sensing elements within the chamber should be
shielded to prevent their contact with or receipt of, direct radiation
from a heating or cooling surface.

Test items having exposed connectors, switches, relays, etc., are frequently subjected to temperature chamber tests. If a functional test of
the item is required, condensation of moisture or the formation of frost
on the contacts of electrical parts could cause a malfunction. This problem is particularly severe during temperature shock tests. If the intent
of the test is to determine whether these electrical contacts will function
after exposure to a particular temperature environment, special devices
to protect them from the ambient environments should not be used.
However, if the electrical parts normally function within some type of
enclosure or are protected in such a manner that conditions causing a
malfunction would not be encountered, special protection for the
exposed contacts should be provided.

For example, connectors should be protected by a metal or plastic
cover expressly designed to fit the exposed end. Switches and relays
may be protected by a plastic bag; however, protecting the entire test
item from the chamber environment by enclosing it in a plastic bag or
similar cover is not permitted unless the arrangement simulates the
expected use conditions. The bag introduces a substantial temperature
time lag. When potted components are exposed to a thermal shock test,
a bag enclosure will prevent moisture from entering cracks that may be
present in the potting compound, thus preventing a failure that would
ordinarily occur in an unprotected item.

6. *Temperature Shock Tests.* Although most of the difficulties
encountered in normal-temperature chamber tests also apply to temperature shock test, some additional considerations apply specifically to temperature shock tests. A normal high-temperature or low-temperature
test, either cycling or constant temperature, is conducted in a single
chamber, since temperature changes are relatively slow and normally do
not exceed the capability of the refrigeration or heating system. For a
temperature shock test, however, it is difficult to attain the desired temperature changes with a single chamber within the specified time. The
standard procedure is to use two separate chambers, one stabilized at
one of the temperature extremes and the other stabilized at the other
extreme. The item under test is then moved from one chamber to the
other within the time called for in the detailed test specification.

Even when two chambers are used to perform a temperature shock test, it is difficult or impossible to maintain a constant or even chamber temperature after a relatively large test item has been introduced into the chamber. For example, if a large item has been stabilized at the low-temperature extreme, it is difficult for the chamber to maintain the required high temperature. This problem can be minimized by using chambers with a volume much larger than that of the test item or with a high capacity for heating or cooling.

7. *Use of CO_2 and N_2 as Coolants.* Liquid carbon dioxide (CO_2) and liquid nitrogen (N_2) are frequently used as coolants in environmental equipment. Often these liquids are injected directly into the space to be cooled. Such a system should be used judiciously, and certain precautions must be observed.

High-pressure liquid CO_2, when allowed to expand rapidly, as it does in injection systems, will partially sublime to the solid form. The resultant "snow" will tend to accumulate in areas of slow air movement or will adhere to surfaces. Thus the "snow" may adhere to a temperature sensor or to test items, subjecting them to a temperature of $-110°$F and thereby causing unwanted effects.

It is desirable to have all the CO_2 and N_2 in the gaseous phase before it enters the chamber volume where test items are located. Injecting CO_2 into the circulated air stream and using baffles and/or a manifold with many small jets should be considered as possible safeguards against the accumulation of CO_2 "snow" on test items and/or temperature sensors. "Snow" with N_2 is normally not considered to be a problem.

Data Interpretation

In general, temperature recording and interpretation are straightforward.

HUMIDITY TESTS

General guidance and suggested minimum requirements for conducting humidity tests within the temperature range of 60 to 160°F and relative humidity range of 15 to 100% are given. Humidity chamber testing consists of subjecting one or more test items to either varying or constant humidity or constant temperatures, as specified in detailed test specifications.

The moisture, or humidity, within the chamber is normally expressed in terms of "percent relative humidity." The relative humidity content of a chamber is usually determined by measuring the dry-bulb temperature of the air and concurrently obtaining a measurement of the temperature differential between the dry and wet bulbs. Using these two measurements, the percent-relative-humidity can then be ascertained from prepared tables. These measurements must be made as accurately as possible, since small errors in these measurements can result in significant errors in the percent-relative-humidity determination. To minimize instrument errors, the dry-bulb and wet-bulb sensing devices should be matched so that inherent errors in both devices are in the same direction and are approximately the same over the specified temperature range.

Most of the difficulties encountered and the precautions taken in performing a proper humidity test are the same as those required for temperature testing. In addition, it is common practice to protect uncoupled electrical connectors from exposure to humidity by sealing the connector with a mating cover or connector during the humidity test. When functional tests are required during the humidity cycle, test cables extending from the test item to the chamber exterior should be provided. If this is impractical, the test item must be removed from the chamber, and dry air (at room temperature) may be used to remove moisture from connectors prior to further testing.

In general, any protection to sensitive parts should simulate the protection provided in actual end use. Also, consideration should be given to the position or orientation of the test item in the chamber to minimize the collection of condensate in depressions or wells.

Test Specification Check List

Certain information is required to assure a properly conducted constant humidity test. The following points should be listed or considered in the preparation of the test specifications:

1. Dry-bulb temperature level(s).
2. Dry-bulb temperature tolerance. A reasonable absolute tolerance is $\pm 10°F$ within the range of 60 to 160°F.
3. Rate of dry-bulb change.
4. Percent relative humidity and tolerance. A tolerance of $\pm 5\%$ is reasonable for a test conducted at a constant dry-bulb temperature. Tests involving a cycling of dry-bulb temperature require chambers equipped with wet-bulb depression measurement and control systems in order to control humidity to $\pm 5\%$ during the constant dry-bulb portion and $\pm 8\%$ during the changing dry-bulb portion of the cycle. The less desirable independent dry-bulb-wet-bulb systems can be used for tests involving cycling of dry-bulb temperature where the humidity tolerances are $\pm 10\%$ and $\pm 16\%$, respectively.
5. Functional tests. Under what conditions and time intervals are such tests to be performed? Specify tests. Normally, functional tests should be performed only after completion of the humidity tests.
6. Special protection to sensitive parts.

Instrumentation

1. *Two Methods Used to Obtain Relative Humidity Measurements.* Humidity chambers are usually equipped with instrumentation connected in such a manner that measurement records of either dry- and wet-bulb temperatures or the measurement of dry-bulb temperature and the amount of wet-bulb depression is continuously recorded. Either of these methods can be used to control or determine the percent of relative humidity in a chamber at any given time. However, instrumentation that will record dry-bulb temperature and wet-bulb depression (Method B, Figure 4) has several advantages over the instrumentation that records dry- and wet-bulb temperatures independently (Method A, Figure 4). The advantages of using Method B over Method A are discussed in the following paragraphs.

Method A shows a scheme of the system used when recording independent measurements of both dry- and wet-bulb temperatures. Method B shows a scheme of the system used when recording the dry-bulb (DB) temperature and the wet-bulb (WB) depression. Method B is preferable to Method A, since there are fewer inherent errors affecting relative humidity determination.

The sources of error significantly influencing relative humidity determination in both Methods A and B are as follows.

Method A

a. Instrument errors: DB sensing, WB sensing, DB indicating or recording (readout system), and WB indicating or recording (readout system).
b. Human errors: DB reading, WB reading.
c. WB scale.

Method B

a. Instrument errors: DB sensing, WB sensing, and differential recording (readout system).
b. Human error: Differential reading (WB depression).
c. Differential scale.

The differential measurement technique has certain advantages over separate measurements of dry-bulb and wet-bulb temperatures. The wet-bulb depression system requires only one instrument to determine the humidity, whereas the independent system requires two. Since there is only one instrument, errors in the dry-bulb and wet-bulb sensing devices can be balanced out by zeroing the readout system, while both temperature sensing devices are immersed in the same temperature medium. Thus, all three types of instrument errors are minimized at the calibration temperature and over a range extending above and below this temperature. (This may be demonstrated by repeating the calibra-

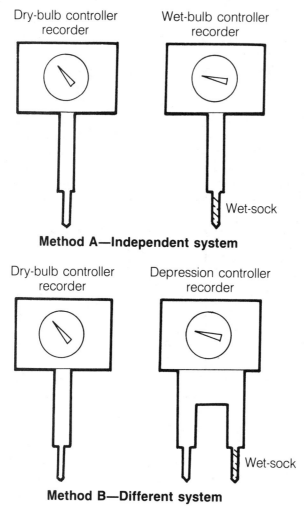

Figure 4 Recording methods.

tion or balancing procedure at other temperatures above and below the first one performed.)

Since the accuracy of the readout of most control instruments is rated as a percentage of full-scale (0.25%, for example), a small range is desirable. Small ranges with expanded scales are readily available for use with a depression system. Ranges of 0 to 10°F depression are quite common and 0 to 30°F are also used. These scales on a 0.25% instrument would yield readout accuracies of ±0.025 degree and ±0.075

degree, respectively. (On an independent wet-bulb instrument, a scale of 60 to 160°F would yield a readout error of 0.250 degree. For these reasons, small ranges with expanded scales are desirable.)

Finally, sources for human error for RH determination are reduced by 50% because only one reading is needed instead of two.

2. *Example of Importance of Accurate Temperature Differential Measurements.* When wet-bulb depression is obtained by an indirect method (Method A), such as subtracting an independent wet-bulb reading from another independent dry-bulb reading, large errors are possible in determining the percent relative humidity. Example 1 (below) illustrates how a small error in the separate measurement of dry-bulb and wet-bulb temperatures can result in a significant error in relative humidity determination. Example 2 shows the advantage of using a direct temperature-differential measurement method.

For the purpose of these examples, the actual conditions of temperature and humidity within the test chamber are to be considered as follows:

Dry-bulb temperature = 68°F.

Wet-bulb temperature = 67°F (or temperature differential 1 = 1°F).

Relative humidity (from Table 1) = 95%.

Example 1. Assume that the wet-bulb temperature is 67°F, but that the dry-bulb temperature indication is 1°F too high (69°F instead of 68°F). The temperature differential will then appear to be 2°F (69°F minus 67°F). From Table 1, it can be seen that the relative humidity will appear to be 90% or five percentage points less than the actual condition. On the other hand, if the dry-bulb temperature is 1°F too low (67°F), the temperature differential will appear to be 0°F and the relative humidity 100% or five percentage points too high. If the test specification has a relative humidity tolerance of plus or minus five percentage points, in both cases the measurement error will be equivalent to the test tolerance. Notice that the same relative humidity errors will be present if the dry-bulb temperature is correct, but errors of ±1°F occur in the wet-bulb temperature. However, if the dry-bulb temperature is 1°F too high (69°F) and the wet-bulb temperature 1°F too low (66°F), the indicated temperature differential of 3°F will give a humidity reading of 85% or an error of ten percentage points.

Example 2. Assume the test conditions of Example 1, and use a direct temperature-differential measuring device to measure differential. A 1°F error in measuring the dry-bulb temperature (either 69 or 67°F) will not result in an erroneous relative humidity reading because zeroing the readout system balances the instrument errors in the temperature differential measuring device during the calibration procedure. Conse-

Table 1 Percent Relative Humidity from Wet- and Dry-Bulb Temperature Measurements (29.92 Inches of Mercury)[a]

$t-t'$[b]	0.2°	0.4°	0.5°	0.7°	1.0°	1.5°	2.0°	2.5°	3.0°	3.5°	4.0°	4.5°	5.0°	6.0°	7.0°
36°	98	97	95	93	91	86	81	77	72	65	63	60	55	48	38
50°	99	98	97	95	93	90	87	83	80	78	74	71	68	62	56
68°	99	98	97	96	95	93	90	88	85	83	81	78	76	71	67
86°	99	99	98	97	96	94	92	90	88	87	85	83	81	77	73
100°	99	99	98	97	96	95	93	91	89	88	86	84	83	80	77
110°	99	99	98	98	97	95	94	92	90	88	87	86	84	81	78
120°	99	99	99	98	97	95	94	92	91	88	88	86	85	82	80
148°	99	99	99	99	98	97	95	94	93	91	90	88	87	85	82
160°	99	99	99	99	98	97	95	94	93	91	90	89	88	85	83
180°	99	99	99	99	98	97	96	95	94	93	92	90	89	87	85
198°	99	99	99	99	98	97	96	95	94	93	92	91	90	88	86

[a]This table is a portion of a relative humidity extracted from a larger table commonly used to determine relative humidities.
[b]t = dry-bulb termperature, °F; t' = wet-bulb temperature, °F.

quently, a 1°F temperature differential is recorded as existing in the chamber yielding the accurate relative humidity reading of 95% (Table 1).

Test Operation and Equipment

1. *Differential Temperature Measurement.* The best method to obtain acceptable limits of relative humidity in a chamber is to provide the chamber with controls that will keep the temperature differential within the specified limits. To do this, the instrumentation and sensors must be selected to provide the best possible control accuracy of the temperature differential.

Careful selection of the sensors is required to minimize the difference error. These sensors must be selected in matched pairs and must indicate the same temperature within very close tolerances when exposed to the same environment (both sensors dry and with the wick removed). The sensors should be checked at the extremes and at the mid-point of the temperature range.

2. *Independent Wet-Bulb and Dry-Bulb Measurements.* The major parameter in humidity control is the temperature difference between the wet-bulb and dry-bulb sensors. It is imperative, therefore, that the errors in the measurements be controlled in such a manner that the temperature difference be known as accurately as possible. A small error in the actual temperature measurement of the wet-bulb or dry-bulb sensors is not as important as the difference between the measurements.

Cams for controlling the two temperatures must be accurate, since small errors in these may offset the above-mentioned precautions. This is especially true at the lower temperature extremes where the allowable temperature depression range is smallest.

3. *Wet-Bulb Wick and Air Flow.* Usually, the covering of the wet-bulb sensor is a muslin material used as a wick to carry water from a reservoir to the sensor. Since the rate of water evaporation from this wick determines the resultant temperature measure, certain minimal precautions should be observed. Salts and other water contaminants can accumulate on the wick and restrict or impede the capillary action; thus, only distilled (preferred) or deionized water should be used. Even atmospheric dust will be accumulated and will likewise affect the water supply. It is then good practice to change the wet-bulb covering periodically—at least every ten days of operation.

Air flow past the wet-bulb sensor is also important and should be in the range of 400 to 1200 ft per minute, with a suggested velocity of 900 ft per minute. If both the dry-bulb and wet-bulb sensors are located in the same airstream, the dry-bulb sensor should be located beside or upstream from the wet-bulb sensor. This minimizes the cooling effect on the dry-bulb sensor of the evaporating water from the wet-bulb wick. The wet-bulb and dry-bulb sensors should be separated to minimize any radiant heating or colling effects between the two sensors.

4. *Chambers.* The remarks under Temperature Tests also apply to humidity chambers. (See pages 152−153).

An additional precaution must be observed. It is necessary to prevent drippage of condensed moisture from falling onto the test items. This is most likely to occur during the increasing dry-bulb temperature portion of a constant humidity test. At this time, the chamber wall temperature, which lags the air temperature, will frequently be below the dew point of the chamber atmosphere. Thus, it may be necessary to install a baffle and provide for drainage. A ceiling with a slope sufficient to allow the condensed moisture to run down the walls is an alternative solution.

The method of supplying moisture should receive attention. Heated water baths have the disadvantage of being slow to react. This can often cause a severe cycling of wet-bulb indication. The injection of low-pressure steam has been successfully used and, when used in conjunction with differential method for control, very close tolerances have been achieved.

Data Interpretation

Although humidity data recording and interpretation are straightforward, a few precautions should be observed:

1. The dry-bulb and wet-bulb data should be so designated. If separate recordings are made, they should be indexed or keyed to the same time scale so that related data may be correlated.
2. The average annual station pressure (mean or average to the nearest inch, or equivalent centimeter, of mercury) may be obtained from the nearest weather bureau office and referenced for relative humidity determinations from psychrometric tables.
3. If the wet-bulb temperature depression is reported, the means of determination should be specified (separate wet-bulb and dry-bulb measurements, or the temperature differential measurement).

MECHANICAL SHOCK TESTS

The mechanical shock test is one of the most difficult environmental tests to conduct in such a way that the results can be correlated with those of different agencies. The difficulty stems partially from the fact that the effect of a pulse is to excite resonant vibrations in the test item.

A linear single-degree-of-freedom system may respond at acceleration amplitudes up to twice the acceleration amplitude of the pulse, depending on the relationships between the applied pulse shape and duration and the resonant frequency of the test item. If the test item is more properly regarded as a nonlinear multidegree-of-freedom system, or if the test pulse has a superimposed vibration on it, the response ratio may be much greater than two.

Determining the shock pulse to be applied is not a simple matter, since many factors are involved. Shock pulses may be characterized by such parameters as amplitude, duration, rise and fall times, velocity change, and pulse shape. In some cases these parameters can vary widely without any appreciable effect on the test item. In such situations, only the critical parameter(s) need to be closely controlled. The designer should have some insight into the effect of changes in the parametric values if he is to have an appreciation of the survival limits of his component to mechanical shock. For example, to speak of an item as being "good to 5000 g" is meaningless unless one also describes some other characteristics of the shock pulse, such as duration or rise time. It is possible for an item to fail from a 500 g, 10 ms half sine pulse and yet survive a 5000 g, 0.3 ms shock. Sometimes the shock-test parameters can be specified without much knowledge of the test-item structure when data is available that describes the use environment.

However, to be meaningful, such data must have been obtained at the attachment points of the component to the structure, with the component in place. If structural changes are made in the carrying system, the component may be subjected to a different shock pulse, with the input to the system unchanged. In the absence of "use environment" data, estimates of the component shock parameters can often be made, based on system mathematical models and input shock. As field data becomes available, better estimates of the appropriate shock test for the component can be made.

The most frequent problems noted in mechanical shock testing are inadequate specifications, poor instrumentation techniques, inferior holding fixture design, generated pulse deviations, and improper evaluation of test data.

Now we shall discuss techniques that will alleviate the severity of these problems.

Test Specification Check List

Certain information is needed to assure a properly conducted mechanical shock test. The following items should be listed or considered in the preparation of the text specifications.

1. *Applied Shock.* The applied shock can be characterized by many parameters: maximum faired acceleration (amplitude), duration, rise and fall times, and velocity change. The importance of each parameter depends on the test-item design and the purpose of the test. First, one must choose the parameters to specify, and then the tolerances necessary to control the test pulse. In general, a shock that significantly excites a critical frequency of the test item is potentially the more damaging. The relationship between the pulse duration, rise and fall times (pulse shape), and the critical frequency(ies) of a test item is very significant for, depending upon their relative values, the amplitude response of the test item may be much smaller or much greater than the amplitude of the applied pulse. In the simple case where the test item is a linear single-degree-of-freedom system with a viscous damping ratio < 0.1, the pulse duration that excites the critical frequency of the test item most severely can be estimated as follows:

$$D = T = \frac{1}{f_c}$$

where

$$
\begin{aligned}
D &= \text{pulse duration, ms} \\
T &= \text{period of critical frequency, ms} \\
f_c &= \text{critical frequency,}[5] \text{ kcps.}
\end{aligned}
$$

The above formula can be used with an error in response amplitude of less than 4% when the test pulse is symmetrical. If a nonsymmetrical pulse is used, the desired duration can be determined from the normalized shock spectrum of the input pulse shape.

The shock applied to a test item should be specified in one of three different ways [see paragraphs 1(a) to 1(c) below] when a linear single-degree system with a damping ratio less than 0.2 is assumed. The particular selection depends on the relationship between the applied pulse duration D and critical period T. If the test item cannot be regarded as a simple linear system, or if critical frequencies are unknown, then the applied shock should be specified, using paragraph (b), below. If the applied shock is desired as that which produces a given shock spectrum (with limits), the spectrum should be converted[6] to some particular input pulse as specified in paragraph (b), below.

a. $D \leq 0.2T$. Specify velocity change, using the tolerances recommended in paragraph 1(d)(4)(a) and maximum duration only. Maximum faired amplitude, rise and fall times, and pulse shape are not important and should not be specified.

b. $0.2 < D < 5T$. Specify maximum faired amplitude, duration, rise and fall times,[7] and velocity change, using the tolerances suggested in paragraph 1(d).

[*Note*: It is permissible to specify only maximum faired amplitude, one other parameter, and a pulse shape; see paragraph 1(d)(5).]

Characteristic parameters (duration, rise and fall times, velocity change) of several classical shapes can be determined from the mechanical shock nomograph (Figure 5), but a haversine shape is the easiest to generate.

c. $D \geq 5T$. Specify minimum duration, maximum faired amplitude, and rise and/or fall time (if the rise and/or fall times are less than $0.3D$) using the suggested tolerances of paragraphs 1(d)(1) and 1(d)(3).

d. Tolerances. When the applicable parameters are specified, the following absolute tolerances should be used:

(1) Maximum faired amplitude, ±15%
(2) Duration:

    (a) 2 ms and over, ±15%, whichever is greater.
    (b) Less than 2 ms, ±0.1 ms or ±30%, whichever is greater.

(3) Rise and fall time:

    (a) Over 1 ms, ±0.3 ms, or ±15%, whichever is greater.
    (b) 0.05 to 1.0 ms, ±30%.

(4) Velocity change:

    (a) +15% [for paragraph 1(a)].
    (b) +25% [for paragraph 1(b)].

(5) Pulse shape, tolerance, not specified per se, so shape must be translated into equivalent duration, rise and fall times, and velocity change; for tolerance see paragraphs 1(d)(2) to 1(d)(4).

2. *Additional Requirements.*

a. Definition of axes and directions of applied shock force (shock acceleration).
b. Number of shocks in each direction.
c. Test sequence of directions, when important.
d. Point of application, if different from that through the normal mounting features.

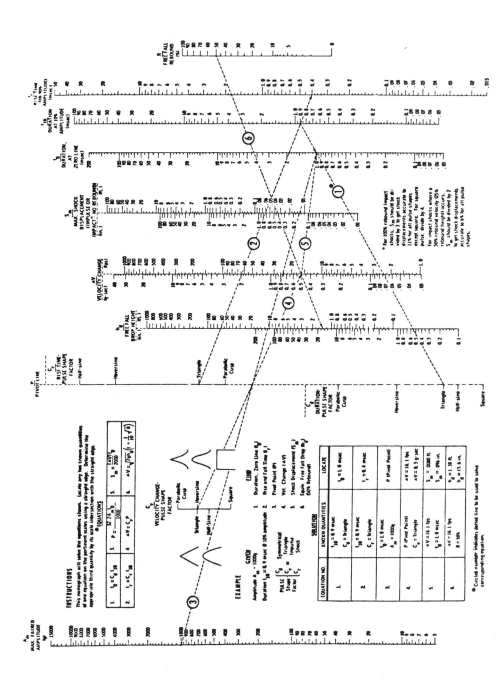

3. *Miscellaneous.* The following items need not always be specified but should be considered.

a. Limits on "preshock" or "postshock" accelerations. (The effects of successive pulses should be considered.)
b. Location, orientation, and maximum allowable weight of transducers, if mounted on test item to measure response.
c. Test-item temperature if other than room temperature.
d. Upper limit of frequencies of concern (to determine limits for fairing).
e. Test-item mounting requirements.

   (1) If the "normal mounting features" are not readily apparent, describe the means to be used (potted, strapped, etc.).
   (2) The need for special mounting bolts or screws.
   (3) Fastener torque.
   (4) The kind of washers to be used, if any.
   (5) The method to secure or loop excessively long cables.

f. Conditions, intervals, and types of functional testing or monitoring that is to be done on the test item.

4. *Instrumentation.* It is essential that shock tests be properly instrumented. Accelerometers presently applicable to shock measurement are mass-spring systems whose response characteristics are related to the parameters of the applied pulse. The accelerometers should be carefully chosen in order to get a true record of the applied pulse. The associated electrical circuits for each type of accelerometer should also be carefully selected. Block diagrams showing common parts of two shock monitoring systems appear in Figure 6.

5. *Accelerometer Mounting.*

a. Location. Unless otherwise specified in the detail test specifications, accelerometers used to monitor input shock to the item under test should be mounted on the test fixture as close as possible to a test item mounting point, but not touching the test item.

   Accelerometers should not be mounted on the test item to monitor input unless it is impracticable to do otherwise. If they must be mounted on the test item, then care should be taken in locating them. If mounted directly on the test item to monitor response, the weight of the accelerometer and its associated hardware should be as small as practicable.

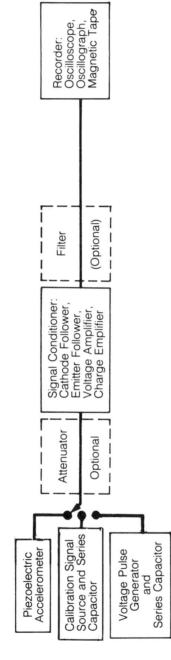

Figure 6 Two types of accelerometer circuitry.

b. Accelerometer Mounting Block. Any accelerometer mounting block used (that is, any auxiliary hardware used as a means of attachment or as a spacer between the accelerometer base and the surface to be monitored) should not change the maximum faired amplitude more than ±2% over the range of shock rise times to be encountered. Metallic blocks are acceptable, but some plastic blocks and electrically nonconducting blocks are not.

c. Surface. The surface used to mount accelerometers should have a surface finish of 16 μin. or better, root mean square, over the area in contact with the accelerometer (and accelerometer mounting block, if used) if the accelerometer (or mounting block) is to be attached with threaded fasteners or cyanoacrylate adhesive. If the accelerometer or mounting block is to be attached with dental cement or epoxy adhesive, the surface may be roughened slightly to ensure proper adhesion. Dental cement or epoxy adhesive attachment (with threaded fasteners for additional strength, if required) should be used for mounting surfaces where the finish of 16 μin. rms cannot be obtained.

The mounting surface should be flat within ±0.0005 in. over the area in contact with the accelerometer and mounting block. The tapped holes used in mounting should be perpendicular within +0.25 degree to the mounting surface.

d. Torque. The torque used in mounting piezoelectric accelerometers should be that specified by the accelerometer manufacturer. Insufficient torque can cause the recorded shock pulse to be distorted.

6. *Accelerometer Selection.* Accelerometers should be either piezoelectric or strain gauge type and should meet the following requirements:

a. Piezoelectric Accelerometer. Some piezoelectric accelerometers have restrictions as to the type of shocks they may receive. These restrictions involve the maximum permissible shock levels in the sensitive and transverse axes, and the stability of calibration.

Any accelerometer used should have the approval of the integrated contractor responsible for procurement. Piezoelectric accelerometers used to measure a shock pulse should have a natural period $(1/f_n)$ less than 0.4 of the expected pulse rise time or pulse fall time.

The natural period restriction is necessary to preclude use of the transducer in a range of shock tests where its transient response ratio is significantly greater than one. If, during a test, the accelerometer is used at a temperature that is not in the range of 35 to 110°F, the accelerometer should be calibrated at the exposure temperature ±10°F.

Most piezoelectric accelerometers exhibit changes in sensitivity dependent upon the amplitude of the accelerations to which they are subjected. If an accelerometer is to be used to measure shock-pulse amplitudes significantly higher than the shock calibration amplitude, a correction factor may be applied to the accelerometer sensitivity. This factor should compensate for the difference in sensitivities at the calibration and test amplitudes, provided that such a factor is available for the particular accelerometer in use. If an accelerometer is to be used to measure very low test pulse amplitudes, calibration using the sensitivity determined by vibration will generally result in greater accuracy than can be obtained by using the shock calibration sensitivity.

b. Fluid-damped Strain Gauge Accelerometer. Fluid-damped strain gauge accelerometers, at the specified test temperatures, should have a damping ratio between 0.4 and 0.8 of critical damping. The accelerometers used to measure a shock pulse should have a natural period $(1/f_n)$ less than 0.8 of the expected pulse rise time or pulse fall time. The natural period restriction is to insure usage of the transducer in a range where the transient response ratio will not differ significantly from one. Fluid-damped accelerometers should not be used when the expected pulse rise time or fall time is less than one millisecond. In addition, the g-rating of the accelerometer should be greater than the specified pulse amplitude by at least 10%. For tests conducted outside the 50 to 110°F temperature range, an appropriate temperature-compensated or temperature-controlled accelerometer should be used.

7. *Attenuator (Piezoelectric Monitoring Circuit Only).* Attenuators (capacitors, additional cable lengths) are sometimes inserted into the circuit to improve the ratio of the time constant of the monitoring circuit to the pulse duration, and to limit the amount of voltage to the input side of the signal conditioner.

8. *Filter.* Sometimes it is necessary to use filters in the readout of piezoelectric accelerometers (fluid-damped strain gauge types are inherently mechanical filters). If filters are required to remove unwanted frequency components from the signal, they should be carefully chosen so that the circuit containing them will meet requirements. The characteristics of the filter should always be known. Some filters suitable for sustained vibration tests are *not* suitable for shock tests. A desirable filter for shock test instrumentation is a transient or Gaussian low-pass filter. This type of filter should have a linear phase shift throughout the frequency range to at least twice the cutoff frequency. (The constant

slope to the phase shift-frequency curve means there is a constant real time delay of all the pulse harmonics at all frequencies to twice the cutoff frequency.) At the cutoff frequency the attenuation of the filter signal should be down 3dB $\pm$ 0.5dB. The selected upper cutoff frequency (f_0) for a filter is dependent upon the shock rise time (t_r) and fall time (t_f), and is found from the following equation:

$$f_0 \text{ (in kcps)} \geq \frac{2.5}{t_r \text{ (in. ms)}} \text{ and } \geq \frac{2.5}{t_f \text{ (in. ms)}}$$

Instrumentation circuits selected to monitor both shock input and response should have identical filters, since filtering in a circuit causes a delay of real time output; for example, a triangle filtered with f_0 = $2.5/t_r$ can cause the recorded peak to lag the true input peak by approximately 15% of the input pulse rise time.

Fixtures

Holding fixtures should mount the item under test by the item's normal mounting features and with suitably torqued fasteners. The holding fixture itself should always be as rigid as practicable. In general, the same fixture design can be used for both shock and vibration tests. However, for shock tests, the period of any fixture resonance should be less than one-half the specified rise or fall time, whichever is shorter. The fixture should be designed so that the specimen can be oriented easily in six different directions. Space should be allowed to mount accelerometers.

Test Operation and Equipment

1. *Shock Pulse Generating Equipment.* Generally, impact machines or impulse machines may be used, provided the pulses so generated can be repeated. It is difficult or impossible to get sufficiently repeatable pulses with impact machines using sand as the impact medium. For the sake of repeatability, all machines should be equipped with an automatic release device. On all machines the "preshock" and "postshock" accelerations should be less than 20% of the maximum faired acceleration of the primary shock, if such secondary accelerations occur before or after the primary pulse at times greater than or equal to $10T$.

It is preferable (and in some cases, mandatory) that impact machines be equipped with a braking mechanism to prevent successive impacts. When impulse machines are used to test certain devices (for example, acceleration sensing devices), special limits may have to be specified on the allowable braking acceleration.

With either impulse or impact machines, the period of the fundamental resonance of the loaded shock-machine carriage should be less than one-half the specified pulse rise time or fall time, whichever is shorter. In addition, the loaded shock machine should be capable of generating pulses 'so that the peak-to-peak amplitude of any superimposed high-frequency ringing located anywhere along the faired amplitude of the unfiltered recorded pulse will be less than ±20% of the maximum faired amplitude.

2. *Test Operation.* Quite often, a requested pulse cannot be generated accurately. If so, the generated pulse should produce at least equal damage potential to the test specimen as the specified pulse would produce. Changes in specified pulse shape, amplitude, duration, rise time, fall time, and velocity change may thus be required. The test item's response to the generated shock should then be considered in these cases.

When there is a significant change in mounting configuration of the test item, the shock pulse should be instrumented. For example, the test item might be shocked 18 times: three times when mounted directly on a plate to the carriage, three times when mounted on a standoff fixture, and 12 times in two axes when mounted on an upright fixture. Usually, when a highly repeatable machine, and a uniform test item are involved, this type of test would require only three shock records, one for each type of fixture setup. If the test item is nonsymmetrical, or the machine is not too repeatable, as many as 18 shock records might be required.

As a check upon a free-fall shock machine and its instrumentation, the drop height and unbraked percentage of rebound should be determined for each acceleration level and duration specified (free-fall heights are measured from the bottom of the carriage anvil to the top of the impact material). The indicated velocity change determined from the shock record should be within ±20% of the velocity change determined from the shock nomograph, knowing the drop and rebound heights.

3. *Nomograph.* The mechanical shock nomograph (Figure 5) can be used to solve a number of different equations frequently encountered in mechanical shock tests in the areas of test specifications, pulse generation, and pulse interpretation.

Data Interpretation

1. *Fairing.* Sometimes on a recorded shock pulse, a difference appears between the "maximum faired amplitude" and "maximum amplitude." When a recorded shock pulse needs to be faired (that is, needs to be smoothed out to determine the basic pulse parameters), then the equipment generating the distorted pulse should be examined to see if the generated pulse can be improved. In many cases the equipment is at fault; in other cases, the equipment is satisfactory but the pulse is still distorted because of feedback from the test item into the shock machine.

If, after all attempts to generate a better pulse have failed and the curve must still be faired in order to determine the pulse parameters, then the following technique should be used.[8]

On a recorded shock pulse signature, draw a smooth line through the higher (unwanted) frequencies so that the amplitudes of the unwanted frequency signals are equal on either side of the fairing line. Stated another way, the fairing line should be drawn through the mid-points of each successive half-cycle (peak-to-peak) of the unwanted vibration signal. If the faired curve is still not sufficiently smooth, the process of fairing out the remaining "wrinkles" can be repeated. This method of fairing will produce a faired pulse equal in area to the recorded pulse as long as the superimposed vibration on the pulse is caused by a linear system.

When an extremely nonlinear vibration occurs on the recorded pulse, the method of fairing will have to be slightly modified. The fairing line is drawn through the higher unwanted frequencies so that equal areas exist on either side of the fairing line.

In any case, the total area of the faired pulse should equal the total area of the recorded pulse.

2. *Recorded Pulse Deviations.* It is good practice to compare all of the pulse parameters, maximum faired amplitude, duration, rise time, fall time, and velocity change of the recorded machine calibration pulse with the same parameters of the requested pulse. Velocity change can be found by measuring the area enclosed by the zero acceleration and faired amplitude lines. The change can also be found by measuring the carriage velocity just prior to and just after the primary pulse. Knowing the instrumentation accuracy and the allowable absolute tolerances, permissible machine operational tolerances can be found. If the overall instrumentation accuracy can be shown to be better than that mentioned in the Process Standard, machine operating tolerances can be increased accordingly.

LOW-PRESSURE TEST

The low-pressure test is conducted to determine the deleterious effects of reduced pressure on aerospace and ground equipment. Damaging effects of low pressure include leakage of gases or fluids from gasket-sealed enclosures and rupture of pressurized containers. Under low-pressure conditions, low-density material tends to sublime and many materials change their physical and chemical properties. Damage due to low pressure may be augmented or accelerated by the contraction, embrittlement, and fluid congealing induced by low temperature. Erratic operation or malfunction of equipment may result from arcing or corona. Greatly decreased efficiency of convection and conduction as heat transfer mechanisms under low-pressure conditions is encountered.

The test procedures described are intended to serve several purposes. Procedure 1 is applicable to ground equipment.[9] The test is conducted to determine the ability of ground equipment to withstand the reduced pressure encountered during shipment by air, and for satisfactory operation under those pressure conditions found at high ground elevations. Procedure 2 is applicable to installed aerospace equipment.[10] This test is performed to determine the ability of equipment to operate satisfactorily following exposure to both reduced pressure and temperature conditions encountered during flight.

Procedure 1. Ground Equipment

The test item should be placed into the test chamber. During the test the internal chamber temperature should be uncontrolled. Internal chamber pressure should be reduced to 3.44 in. of mercury (50,00 ft above sea level) and maintained for a period of not less than one hour. This pressure should then be increased to 20.58 in. of mercury (10,000 ft above sea level), the test item operated, and the results compared with the original data. The chamber should then be returned to room pressure and the test item inspected.

Procedure 2. Aerospace Equipment

The test item should be placed into the test chamber. The test chamber internal temperature should be reduced to $-54°C$ ($-65°F$). The test chamber internal pressure should then be reduced to the lowest pressure condition for which the test item is designed to operate while maintaining the specified temperature. (When performance requirements are specified in altitude in feet, the equivalent pressure can be obtained from the U.S. Standard Atmosphere, 1962.) The conditions of pressure and temperature should be maintained for a period of not less than one hour. After this period and while at the specified pressure and temperature, the test item still operating, the test chamber internal pressure should be gradually increased to room pressure. The rate of pressure change should be as specified in the individual equipment specification. During this period special attention should be given to electrical and electronic test items for erratic operation or malfunction resulting from arcing or corona. The test item should then be removed from the test chamber and inspected.

FUNGUS TEST

The fungus test is conducted to determine the resistance of equipment to fungi. Fungi secrete enzymes that can destroy most organic substances and many of their derivatives. They can also destroy many minerals.

Typical materials that will support fungi and are damaged by them are cotton, wood, linen, cellulose nitrate, regenerated cellulose, leather, paper and cardboard, cork, hair and felts, and lens-coating materials.

Test Operation and Equipment

Four groups of fungi are listed in Table 2. One species of fungus from each group will be used.

In preparation of the spore suspension, sterile distilled water having a pH value between 5.8 and 7.2 at a temperature between 22 and 32°C (72 and 89°F) should be utilized. Approximately 10 ml of sterile distilled water should be introduced directly into each tube culture of the fungus and the fungal spores brought into suspension by vigorously shaking or by gentle rubbing of the spore layer with an inoculating loop without disturbing the agar surface. This process should be repeated for each species of fungus. The separate spore suspensions from the four types of fungi should be mixed to provide a composite suspension.

Table 2

Group	Organism	American Type Culture Collection Number[a]	Quartermaster Number[b]
I	*Chateomium globosum*	6205	459
	Myrothecium verrucaria	9095	460
II	*Memneniella echinata*	9597	1225
	Aspergillus niger	6275	458
III	*Aspergillus flavus*	10836	1223
	Aspergillus terreus	10690	82
IV	*Penicillium citrinum*	9849	1226
	Penicillium ochrochloron	9112	477

[a]Source. American Type Culture Collection, 2112 M Street, N.W., Washington 6, D.C.
[b]Source. Mycology Laboratory, PRD, Quartermaster Research and Engineering Center, Natick, Massachusetts

Actively growing cultures between 7 to 21 days old after initial inoculation should be used for the preparation of the spore suspension. After preparation, the spore suspension should not be kept for more than a 24-hour period at temperatures from 22 to 32°C (72 to 89°F), or not more than 48 hours at temperatures from 2 to 7°C (35 to 45°F).

Procedure 1

This procedure should be used for complete assemblies or large pieces of material which cannot be cut or reduced to sample size. The test item should be placed in a mold chamber, equal to that specified in MIL-C-9452, and installed as specified. The internal test chamber temperature should be raised to 30 ± 2°C (86 ± 3.6°F) at 95 ± 5% relative humidity and maintained throughout the test period. The test item should be sprayed with the suspension of mixed spores. To insure viability of the organism, a known nutrient material, inoculated with the same spore suspension used to spray the test item, should be placed in the test chamber. The test period should not be less than 28 days. At the end of this period the test item should be removed from the test chamber and inspected. If so specified in the individual equipment specification, the test item should be operated and the results compared with those previously obtained.

Procedure 2

This procedure should be used for materials that can be cut or reduced to a size suitable for testing in a petri dish. The test should be performed in accordance with Specification MIL-F-8261, except that the spore suspension should be prepared as specified in paragraph two under the heading Test Operation and Equipment. A test period of not less than 14 days should be used.

SUMMARY

We have presented the philosophy necessary for a successful environmental test program. Detailed specifications, required instrumentation, test methods (with examples), and data collection and its interpretation have been given. Recommended references for additional readings have been listed. The environments covered were temperature, humidity, mechanical shock, pressure, and fungus.

RECOMMENDED READINGS

Temperature Tests

National Bureau of Standards. *Precision Measurement and Calibration.* NBS Handbook 77, Vol. II. Fundamentals of the theory and use of thermocouples may be found on page 68 of this reference. Page 90 lists methods for selecting the thermocouple materials, and page 200 lists publications on temperature measurements.

Beckwith and Buck. *Mechanical Measurements.* Reading, Mass.: Addison-Wesley, 1961. This is a good single reference on the subject. Chapters 1 to 3 present a general discussion of measurements including basic accuracy of measurements. Chapter 13 covers, in detail, the subject of temperature measurements.

Atomic Energy Commission. *ALO Standardization Program Handbook.* AL Appendix 72XF. Albuquerque Operations Office, June 1, 1961. Pages 7 and 8 give definitions for primary, secondary, and transfer standards.

Humidity Tests

These references are recommended in addition to those listed under the heading "Temperature Tests."

Gregory and Rouke. *Hygrometry.* Crosby Lockwood and Son, Ltd., 1957. Chapter 1 presents a very complete introduction to humidity measurements and a review of equipment and techniques. The remaining chapters examine various aspects, such as humidity measurement of high-pressure gases, in greater detail.

Marvin, C.F. *Psychrometric Tables.* U.S. Department of Commerce, W.B. 235. Tables VI through X may be used for relative humidity determinations in the range of −40 to +140°F. These tables are prepared for barometric pressures from 23.0 to 30.0 in. of mercury.

Mechanical Shock Tests

Harris, C.M. and C.E. Creed. *Shock and Vibration Handbook.* Vols. I-III. New York: McGraw-Hill, 1961.
Shock Instrumentation, Chapters 12, 14, 16-19.
Shock Motion Response, Shock Spectra, Chapter 8.
Shock Test Specifications, Chapter 24.
Data Analysis, Chapter 23.
Shock Machines, Chapter 26.
Shock Isolation, Packaging, Chapters 31,41.
Transportation Shocks, Chapters 45-47.

Jacobsen, L.S. and R.S. Ayre. *Engineering Vibrations.* Shock Response Theory, Chapters 1, 3, 4. New York: McGraw-Hill, 1958.

Brooks, R.O. *Shock Testing Methods.* Sandia Corporation Technical Memorandum, SCTM 172A-62(73). September 1963.
Shock Testing Nomenclature.
Problems in Test Specifications.

Shock Pulse Generation.
Shock Test Instrumentation.
Pulse Interpretation.
Normalized Shock Spectra for Classical Pulse Shapes.
B. Epstein, *Amplitude Linearity of Endevco Accelerometers.* Endevco
Corporation Tech-Data Bulletin 641. April 1964.

General Readings

Proceedings Annual Reliability and Maintainability Symposium, 1984:
Environmental Stress Screening—Lessons Learned, Ronald J. DeCristoforo.
Random Vibration Screening to Improve Hardware, Duane R. Dura, John W. Woodward.
Vibrations, Random Required, William G. Kindig, John D. McGrath.

NOTES

[1]See Appendix A, Standards & Calibration.

[2]Temperature shock tests differ from normal-temperature chamber tests in that the test item is rapidly subjected to extreme temperature changes during the test.

[3]Exposure time is the period from the first introduction of the test item into the environment to its removal.

[4]Stabilization time is the time required for the temperature of a test item placed in a temperature environment to reach within $2°F$ of the chamber test temperature.

[5]Critical frequencies f_c are found by vibrating the test item in each of three orientations in accordance with the methods indicated in "Sinusoidal Vibration Tests" in Chapter 8.

[6]Techniques for conversion can be found in the *Recommended Readings*, paragraph 3(f).

[7]Do not specify zero rise (fall) time, as this is impossible to achieve.

[8]In general, fairing should *not* be performed on pulses where the period of the predominant superimposed frequency is greater than one-third of the pulse duration.

[9]Aircraft and missile support. Equipment used outdoors on airfields and missile launching pads for servicing, maintenance support, checkout, etc. Electronic equipment is not included. Communication and electronic equipment of all types and equipment with electronic circuits indoor or outdoor.

[10]Aerospace equipment. Equipment installed in airplanes, helicopters, air-launched and ground-launched missiles.

a. Auxiliary power plants and power plant accessories. (Primary power plants excluded.)
b. Liquid systems. Liquid-carrying or hydraulic-actuated equipment.
c. Gas systems. Gas-carrying or gas-actuated equipment.
d. Electrical equipment. All electrical equipment but not electronic.
e. Mechanical equipment. Equipment having only mechanical operating parts.
f. Autopilots, gyros, and guidance equipment, including accessories, but not electronics.
g. Instruments including indicators, electric meters, signal devices, etc., but not electronics.
h. Armament. Guns, bombing and rocket equipment, but not electronic.
i. Photographic equipment. All aerospace still and motion picture cameras and optical devices.
j. Electronic and communications equipment. All such equipment.

Chapter 8
Vibration and
Acceleration Environments

This chapter discusses in detail the reasons for imposing some requirements that might seem unnecessary. Also we discuss many factors that can have a pronounced effect on the quality and standardization of vibration and acceleration tests, but that cannot be defined in precise requirement form. Therefore, this information deserves attention. In addition, a precise method for measuring and controlling input acceleration amplitudes of sinusoidal signals that are actually complex has never been accepted nationally by engineers engaged in vibration testing. As a result, comparison of vibration test effects and test data between tests controlled by different methods, or even by the same method, has often been impossible. By following the test methods described in this chapter, repeatability and uniformity can be achieved.

SINUSOIDAL VIBRATION TESTS

Sinusoidal vibration test specifications, at the present time, intend that the test item experience (along one axis at a time) pure translatory and sinusoidal input motion at specific frequencies and acceleration or displacement levels. The intended tests are never achieved because structural materials do not have infinite stiffness, vibration systems do not have power supplies with perfect sinusoidal fidelity or infinite power capacity, mounted transducers do not have perfect response characteristics or zero mass, and test fixtures do not have infinite mass or stiffness. Also, the dynamic behavior of the test specimen, when vibrated, can cause mechanical feedback to the input control transducers that results in a distorted input signal.

If the amplitude of the distortion is very large, wide variations in the measurement of the input signal will result from the various methods and types of meters used to measure acceleration. However, by eliminating many of the variables, and by defining some of the commonly misunderstood terms, better correlation of test results is possible.

Test Specification Check List

Certain information is required to assure a properly conducted sinusoidal vibration test. The following points should be listed or considered in the preparation of the test specification:

Mounting

The following items must specified:

1. Definition of axes.
2. Location of input control transducers.

The following items need not always be specified, but should be considered:

3. Are special mounting bolts or screws needed?
4. Are washers to be used? What kind?
5. What torque is to be used on mounting bolts or screws?
6. Are normal mounting means to be used?
7. When the item does not have normal mounting means, should the item be supported in some special manner? Specify in detail.
8. If the item has cable outputs, are these to be tied or looped?
9. If response of the item is to be determined, where should transducers be located?
10. Will the mass of accelerometers and associated hardware be significant enough to affect the item's response?

Critical Frequency Search

The following items must be specified:

1. Input amplitude levels (displacement and/or acceleration; an absolute tolerance of ±15% is reasonable).
2. Frequency range and frequency tolerance (an absolute tolerance of ±2% or 3 cycles, whichever is greater, is reasonable).
3. Definition of critical frequency, and how it is to be determined.

The following items need not always be specified, but should be considered:

4. Will the time of survey be subtracted from cycling time?
5. Is it necessary to obtain sufficient data to plot a complete spectrum or the response of the test item, or will a definition of the maximum response or critical frequency suffice?
6. Is the sequence of tests important? Must surveys of all axes be completed before any other vibration test? [*Note*: Because fewer fixture changes are required, time is saved in the performance of tests if all surveys (vibration at critical frequency) and cycling are completed in each axis before starting on the next axis.]
7. If critical frequency is defined as maximum ratio of the output fundamental to the input fundamental, what ratios are to be considered significant?

Vibration at Critical Frequency

The following itms must be specified:

1. Input amplitude (displacement and/or acceleration).
2. Time at critical frequency.
3. If significant critical frequencies are not found (see Critical Frequency Search item 7) which course of action is to be taken? Extra cycling time? Use some fixed frequency?

Vibration Cycling

The following items must be specified:

1. Input amplitude (displacement and/or acceleration).
2. Frequency range and frequency tolerance (an absolute tolerance of ±2% or 3 cycles, whichever is greater, is reasonable).

Instrumentation

To attain the overall instrumentation system accuracies required in sinusoidal vibration tests, the following characteristics and accuracies of the individual circuit elements and overall circuit schematics, where applicable, should be considered as limits throughout the test range.

Circuit Elements

1. Piezoelectric Accelerometers.

   a. Frequency response flat within ±5% from 5 cps to 3 kc.

2. Velocity Transducers.

   a. Frequency response flat within ±5% from 10 to 100 cps.
   b. Amplitude linearity of ±2%.
   c. Cross-axis sensitivity, if applicable, of less than 5%.

3. True RMS voltmeters (used for acceleration measurements)

   a. Measure the true rms value of a complex signal to a maximum crest factor (ratio of peak to rms) or 4.5.
   b. Accuracy of ±3% of reading from 15 to 10,000 cps and ±5% of reading from 5 to 15 cps.

4. Average Responding Voltmeters. Average responding voltmeters used for vibration amplitude measurement (displacement, velocity, or acceleration) should have an accuracy of ±2% of reading from 10 to 10,000 cps.

5. Peak Reading Voltmenters. Peak reading voltmeters used for acceleration measurements should have an accuracy of ±2% of reading from 20 to 10,000 cps, and ±4% of reading from 5 to 20 cps.

6. Cathode Followers and Amplifiers (used in vibration amplitude measuring circuits).

   a. Frequency response flat within ±2% from 10 to 10,000 cps.
   b. Amplitude linearity of ±1%.

7. Low-Pass Filter "A" (for input control circuits when testing up to 500 cps).

   a. Frequency response flat within ±2% between 50 and 500 cps.
   b. Frequency response down at least 2 dB at 800 cps with a rolloff of at least 18 dB per octave.

8. Low-Pass Filter "B" (for input control circuits when testing from 500 to 2000 cps).

   a. Frequency response flat within ±2% between 500 and 2000 cps.
   b. Frequency response down at least 2 dB at 3000 cps with a rolloff of at least 18 dB per octave.

9. Variable Low-Pass Filters.

   a. Amplitude error caused by filter less than ±2%.
   b. Rolloff rate of at least 18 dB per octave.
   c. Manual or automatic frequency tracking.

10. Band-Pass Filters (automatic tracking filter preferred).

   a. Frequency response flat within ±2% from 50 to 2000 cps.
   b. Bandwidth should be narrow enough to attenuate any second harmonic distortion at least 10 dB, but not greater than 100 cps.
   c. Rolloff rate of at least 18 dB per octave.
   d. Manual or automatic frequency tracking.

11. Frequency Indicators. Frequency indicators should measure the fundamental frequency within ±1.5% or ±2 cps, whichever is greater.

Overall Instrumentation

The following schematics describe typical circuits for measuring displacement and acceleration amplitudes for both input control and output or response measurements. Special attention should be given to the overall circuits and to the individual elements of the circuits to ensure proper impedance matching of the circuit and a flat frequency response of the overall circuit.

1. Input Control Circuits. The block diagram shown in Figure 1 can be used for input control in the frequency range up to 500 cps depending on the type of motion control transducer used. The block diagram shown in Figure 2 can be used for input acceleration control in the frequency range up to 2000 cps.
2. Output or Response Circuits. The block diagram shown in Figure 3 can be used for output or response acceleration measurements. Care should be taken to insure proper circuit sensitivity when using variable low-pass or band-pass filters; however, the second harmonic should be attenuated to at least 10 dB. The fundamental frequency signal should not be attenuated, unless the circuit sensitivity has been corrected for any attenuation.

Fixtures

Vibration fixtures should be stiff enough to transmit the desired vibration level from the moving element of the vibration machine to the test item mounting points without excessive attenuation or amplification. It is even more important that all test item mounting points are receiving the same input level from the table. Also, most tests require that the test item shall be excited in three orthogonal axes, which may require a different fixture for each axis.

Four general rules govern the design of vibration test fixtures:

Fixture Weight

Vibration fixtures should be designed to have as large a mass as practicable. Massive fixtures act as high impedance to feedback from the test item, which might otherwise cause distortion of the input signal and excessive transverse motion at the input. Certainly, some compromises must be made in the limitation of the machine; for example, the required acceleration level, size, and mass of the item to be tested are to be considered.

Dynamic Balance

To attain pure translatory motion of the test item, the dynamic center of gravity of the fixture plus test item must be located over the center of gravity of the moving element of the shaker. In most cases, it is difficult (if not impossible) to design a fixture so as to maintain its dynamic center of gravity over the center of gravity of the moving element due to phase shifts of the test item and fixture system as the frequency is varied. Usually the only recourse is to strive for symmetry and to design a fixture with the static center of gravity of the fixture-test item combination over the center of gravity of the moving element.

When a universal upright fixture is used, caution should be exercised in counterbalancing the test item. Counterbalancing with rigid masses does not result in dynamic balance at all frequencies due to phase shift in the motion of the test item relative to the motion of the more rigid counterbalance mass. Sometimes a spare test item is used to counterbalance the actual test item. This approach may be a satisfactory method, provided that the units have similar characteristics that are relatively small.

Fixture Materials and Method of Construction

Aluminum and magnesium are suitable fixture materials. Some advantages may be gained by using magnesium as compared to aluminum alloys due to higher damping properties obtainable from magnesium. However, the primary reason for using magnesium in preference to aluminum is the fact that, with magnesium, a stiffer fixture for a given mass may be designed. Furthermore, on occasion, it is necessary to fabricate a fixture of a given physical size and yet not exceed a specified mass. In such a case, magnesium has a definite advantage over aluminum. Beryllium is a desirable fixture material from the standpoint of its physical properties of high modulus of elasticity-to-density ratio, but is undesirable because of machining problems and high cost.

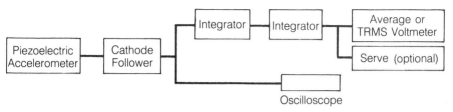

Displacement Control from Accelerometers

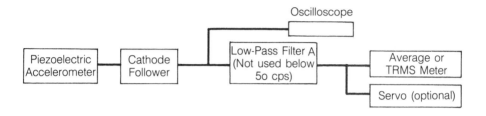

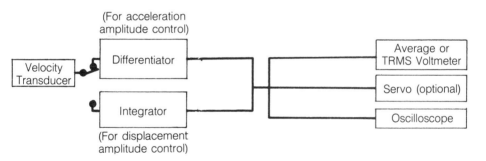

Figure 1 Low-frequency input control circuits.

The above comments should not be construed to violate the concepts of massive fixtures as discussed in the Fixture Weight paragraph. It is frequently necessary to use lightweight materials, such as aluminum and magnesium in the design of fixtures to accomplish such a requirement as physical size consistent with fixture stiffness and yet have a fixture as massive as practicable.

Fixtures fabricated from plate, billet stock, or castings are preferred. Welded and bolted construction should be avoided if at all possible. No general limits can be specified as to the minimum or maximum spacing of the tiedown bolts, since spacing is a function of stiffness of the unsupported structures as well as the load being carried.

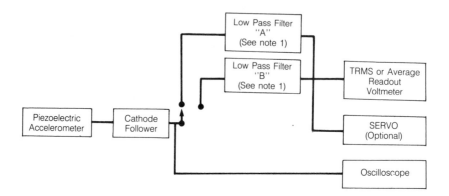

Figure 2 High-frequency input control accelerometer circuit. Note 1: Filters "A" and "B" (par. 1.g and 1.h) are the minimum allowable only. Variable low-pass or band-pass filters are recommended.

Design Frequency

Fixtures should be designed for as high a resonant frequency as possible. Increasing the resonant frequency of a fixture is not always as simple as it first may appear. For example, to double the frequency of a simple spring mass system requires an increase of four in spring stiffness. The acceleration gradient across the height of the fixture is one problem frequently encountered with a vertical type of mounting fixture. As a general rule, try to design a fixture to have a resonance three times the maximum test frequency and the acceleration gradient will be 10% or less.

Test Operation and Equipment

Input Control Point

Unless otherwise specified, control the input to a shaker system above 100 cps from a transducer located on the test fixture immediately adjacent ot a test item mounting point rather than by a built-in transducer located in the moving element of the shaker. The reason for this is that the transmissibility ratio between the two transducers cannot remain one-to-one as the frequency changes.

It is an established fact that radical acceleration gradients will often occur at the various input points around the larger test items, especially at the higher frequencies. Therefore, controlling from a single accelerometer when testing these larger test items (to the order of 6 in. on a side) will not guarantee that the mean acceleration of the test item was

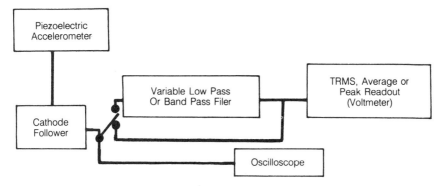

Figure 3 Output acceleration measurement circuit.

as specified. The reasons for these gradients are that the vibration machine armature and fixtures are not infinitely stiff, the unit forces around the periphery of the drive coil are not precisely equal, and the dynamic feedback of the test item will cause various flexing and rocking modes in the armature/fixture. Good vibration machine and fixture design will help to minimize these gradients, but will not solve the problem completely.

If several accelerometers are used to monitor the input accelerations and their average value, derived from an averaging circuit, is used for the input control, at least the mean input motion to the test item would be as intended by the test specification. The use of averaging circuits in the manner described is encouraged.

Mounting Of Transducers

Transducers must be properly mounted so that the signal from the transducer accurately defines the motion or acceleration of the point or surface that it is intended to measure. To ensure that the transmissibility of the connection is adequate, it should be proved by actual test. The sensitivity of many piezoelectric accelerometers may be affected if the mounting surface is not smooth or if the tapped mounting hole is not perpendicular to the surface.

Cycling Test

Vibration systems with electronic amplifier power supplies usually have a logarithmic frequency rate of change (a constant period of time per octave). Another cycling rate (the log-log cycle) has been proposed since, for systems with equal damping, it gives the same number of excursions within the resonant bandwidth regardless of frequency.

Since there are differences in the vibration of components with different cycling rates, one must be selected as a standard. The logarithmic rate has been selected because it tends to excite resonances equally.

Automatic Amplitude Servo Systems

Input control by an automatic amplitude servo system is not only permissible but is encouraged. Control by either a servo system or by manual means is simplified if the input signal is properly filtered.

Surveillance During Test

Persons responsible for the quality and production of an item should always insist that the item undergoing test be observed by properly trained personnel. The behavior of the vibration system, the nature of the unfiltered input signal displayed on an oscilloscope, and the behavior of the item under test should be constantly monitored.

By monitoring the input signal, mechanical loosening or failure of parts in the vibration machine or test item and noise problems in the control circuit can often be detected. Malfunctions in the vibration machine or control system could result in an overtest or an undertest if the system is unattended.

Data Interpretation

Amplitude of Acceleration and Amplitudes of Displacement

Acceleration amplitude measurements in sinusoidal vibration testing are frequently expressed in misleading terms, such as "peak-to-peak." This has led to a surprisingly large amount of misunderstanding in interpreting test results. Therefore, the amplitude of an acceleration signal in sinusoidal vibration testing should always be expressed in terms of zero-to-peak, regardless of the type of meter or detector used (for example, average, TRMS, peak). Double amplitude of displacement or peak-to-peak displacement are terms of convenience only, since displacement measurements are frequently made with optical devices. However, the classic equations defining the characteristics of simple harmonic or sinusoidal motion all define maximum amplitudes (displacement, velocity, acceleration) as zero-to-peak. The instantaneous length (displacement) of the projection (y) of a vector (r) rotating at a constant angular velocity (w) at any point in time (t) or angle ($= wt$) can be defined as follows:

$$y = r\sin wt = \text{(displacement)} \qquad (1)$$

and, by differentiating,

$$\dot{y} = \frac{dy}{dt} = rw\cos wt = v \text{ (velocity).} \tag{2}$$

and, by differentiating once more

$$\ddot{y} = \frac{d^2y}{dt^2} = -rw^2\sin wt = a \text{ (acceleration)} \tag{3}$$

Since all of the terms (displacement, velocity, and acceleration) reach a maximum value when the sine or cosine term is equal to unity, the maximum displacement can be written $y = r$, or one-half the peak-to-peak excursion of the vector projection, and $a = -rw^2$, which is one-half the peak-to-peak amplitude of acceleration.

By substituting in the equation $r = D/2$:

D = double amplitude of displacement or peak-to-peak amplitude of displacement (in w)
w = $2f$
f = frequency (cps)

and dividing both sides by f, acceleration due to gravity (386 in./sec^2) we have:

$$\frac{a}{g} = \frac{4^2f^2D}{2g} = G \tag{4}$$

(G is dimensionless quantity numerically equal to the number of gravitational units)

$$G = 0.0511Df^2 \tag{5}$$

Equation (5) applies only when the signal is pure sinusoidal and is now in the form commonly used in vibration testing work to relate acceleration and displacement. Notice again, for convenience of measurement, that displacement is measured as a peak-to-peak amplitude, but acceleration is measured as the zero-to-peak amplitude.

Input Accelerometer Control Signals

It is the intent of sinusoidal vibration test specifications that the test specimen should be vibrated at the specified acceleration level by controlling the amplitude of the fundamental frequency. Therefore, any distortion of the fundamental signal, regardless of the source, is considered to be extraneous to the desired input amplitude and must not be allowed to affect the measurement of the acceleration amplitude of the fundamental input frequency. The input control signal is frequently distorted by the presence of harmonic or high-frequency "random" distortion. Harmonic distortion on the input control signal is the result of a harmonic-excited resonance of the vibration exciter, fixture, or test item (with resulting feedback) or because of nonlinear spring characteristics in the vibration exciter, fixture, or test item (with resulting feedback). High-frequency "random" distortion is usually generated by the relative motion and resulting impact of parts, or bottoming out of nonlinear spring systems within the vibration machine moving element, fixture, or test item. Frequently, however, this type of distortion will be a repetitive wave (periodic) composed of a multitude of harmonics.

The easiest way to eliminate distortion from the input control signal is to use a filter such as a low-pass filter with a variable-cutoff frequency or a tracking band-pass filter between the input control transducer and the control sensing device (voltmeter or automatic servosystem). These filters should have cutoff characteristics that will significantly attenuate all high frequencies (that is, a clean sinusoidal signal should reach the voltmeter or sensing device). Since both the variable-cutoff and tracking-filter methods are somewhat complex and are moderately inexpensive, passive, low-pass filters should be used in the input control circuit. The low-pass filters are simple to use and give acceptable results on most tests, since the distortion that usually causes the biggest difference (or error) between filtered and unfiltered signals is high-frequency distortion above the cutoff frequency of the low-pass filter. *However, the use of filters in the control circuits does not suggest that the presence and nature of these distortions should be ignored.* Alert test personnel may gain important information by continual observation of the unfiltered signal on an oscilloscope. Below are some examples.

a. *Structural failure and loosening of parts in the test specimen, fixture, or vibration machine.* If the displayed signal has distortion characteristics that radically differ from previous cycles of testing, an investigation should be made immediately to determine the cause. Investigation may disclose a test item failure or prevent an "overtest" of the test item if the distortion is caused by the need of vibration equipment maintenance or better fixture.

b. *Faulty accelerometers or accelerometer cables and connections.* Observing the unfiltered signal can sometimes identify faulty accelerometer cables and connections, especially if the problem is an intermittent open circuit.

c. *Harmonic-excited resonance* of the moving element of the vibration machine or of the test fixture frequently shows up as severe single harmonic distortion on the input control signal. The effect of this phenomenon, if undetected, or not understood, may result in improper conclusions. For example, because of the strange and somewhat noisy behavior some vibration machines exhibit at a particular frequency when this phenomenon occurs, it may be improperly concluded that the test item is at a mode of resonance. Harmonic-excited resonance is discussed in more detail later in this section.

d. *Nonlinear response characteristics* of the test item can cause a multitude of harmonics to be fed back to the input control point, especially when the nonlinear characteristics are due to displacement limiting (or bottoming out).

Transverse Motion

Transverse motion, sometimes called cross talk, is the motion or acceleration orthogonal to the primary or intended axis of motion and is usually expressed in percentage of the intended acceleration. Transverse motion or acceleration indications can result from three sources: (a) the machine and fixture, (b) cross-axis sensitivity of the accelerometer, and (c) reaction and feedback from the test item. Elimination of transverse acceleration generated by the vibration machine is primarily a problem of buying good equipment and keeping it in good condition.

An unloaded machine in good condition should generate unfiltered transverse accelerations less than 20% of the filtered input throughout the frequency range used in testing. If the transverse acceleration from the machine is high, then it is extremely difficult to design a fixture so that the transverse acceleration at the test item input is considered acceptable. The fixture should be designed with a transverse resonance beyond the test range to keep the transverse acceleration within limits. There will occasionally be machine and fixture combinations that exhibit bad transverse acceleration characteristics over narrow frequency ranges that cannot be cured by improving the fixture.

The cross-axis sensitivity of accelerometers causes a false indication of transverse acceleration and is usually inherent in the accelerometer. Therefore, only accelerometers with very low cross-axis sensitivity of accelerometers cause a false indication of transverse acceleration and is usually inherent in the accelerometer. Therefore, only accelerometers with very low cross-axis sensitivities should be purchased and used. Even though a vibration machine could exhibit zero transverse motion on the bare table, and if a fixture of finite mass but infinite stiffness were used to hold a test item which had dynamic properties that varied with frequency, it is still possible for large amounts of transverse motion to be indicated. The reason for this is that the motion of a vibration-machine armature and attached load tends to become "mass controlled" as the frequency increases. The flexure system of some exciters does not offer any guidance constraint at these higher frequencies, since the armature system is now seismic or decoupled in the axial, transverse, and rocking modes. Any dynamic imbalance caused by the test item will therefore be reflected by an increased transverse motion measurement. The complete elimination of this natural feedback phenomenon would require a fixture/armature of infinite mass. It has even been proposed by some people that the system would be more severely tested if the input were constrained to a uniaxial motion. In any event, transverse motion caused by feedback from the normal dynamic behavior of the test item, whether fundamental motion or high-frequency distortion, need not be a compromise of the test. It may be merely an indication of the test-item reaction obeying the laws of physics.

Output or Response Accelerometer Signals

Output signals from transducers mounted on the test item at various points monitor the behavior or response of the test item relative to the motion or acceleration of the input point. Since these output signals are frequently complex (nonsinusoidal), unfiltered amplitude measurements alone rarely provide enough information to properly define the response of the test item for the particular input, nor do they always disclose the most critical frequency. The following paragraphs should help to clarify the meaning of terms frequently used in discussions concerning output signals in sinusoidal vibration tests.

Harmonic-Excited Resonance

We have defined harmonic-excited resonance as the condition that exists when the resonant frequency of a system or structure coincides with, and is excited by, a harmonic of the fundamental input frequency in the input power signal. More specifically, the power supplies of all vibration machines have a finite amount of various harmonics. If a single-degree-of-freedom system is being vibrated at some submultiple of its resonant frequency, the specific power signal harmonic that coincides with the resonant frequency of the single-degree-of-freedom system will be amplified by the mechanical gain (Q) of the system at resonance, with the result that the response signal is composed of the fundamental frequency and the resonant frequency. Obviously, the condition or effect is most evident if the spring system has a large amplification at resonance and if the power supply has significant harmonic content. To ensure that the test item is excited into resonance by an input harmonic, and not the fixture or vibration machine moving element when the nature of the output signal indicates the condition, the unfiltered input control signal should be observed. This observation is to make certain that the particular harmonic does not appear and that the amount of harmonic distortion on the input signal is small compared to the output signal. When all the conditions for causing a harmonic-excited resonance condition are present, the phenomenon will usually be apparent at several submultiples of the resonant frequency. Under these conditions, a mode of resonance in a test item or structure can usually be predicted without actually existing in the test item with the fundamental at its resonant frequency. This will even enable the test engineer to identify a mode of resonance when it is beyond the test frequency range. However, the harmonics in the power signal might often serve as a "helpful" tool in determining the response characteristics of the test item; every effort should be made to minimize power signal distortion. This results in an input that does not conform to the specified intent of having all the input energy at a single frequency.

Critical Frequencies

These are frequencies at which the vibration is most likely to cause structural failure or malfunction of the test item. Resonant frequencies are not necessarily the most critical. Test specifications requiring a resonant frequency survey followed by vibration at resonance, normally should be interpreted as requiring a critical frequency survey followed by vibration at the critical frequency. Resonance determination is useful in development testing to determine natural frequencies for design purposes; resonant frequencies, when apparent, should always be documented, even though not considered to be the most critical.

   1. Detecting Resonant and/or Critical Frequencies. Common methods for detecting critical frequencies are listed below. However, the method to be used should be defined during development testing and included the detailed test specifications.

   a. Resonance Determination. Finding the true resonance of any structure or system, except for simple ones, is sometimes difficult. However, for the purpose of this chapter, a mode of resonance will be defined as that frequency at which the ratio of the filtered output to the filtered input (that is, ratio of fundamental frequency amplitudes only) is a maximum (greater than one-to-one, and the ratio has the characteristics of decreasing if the input frequency is either raised or lowered. This definition is directed towards high Q (little damping) structures that would have high stress at resonance. Resonant frequencies of low Q (highly damped) structures, although more difficult to detect, should be determined if possible. The presence of harmonic distortion, on the output signals at submultiples of a specific higher frequency, will often confirm and even allow a prediction that this higher frequency is a mode of resonance being excited by the harmonic distortion in the power supply.

The significant point is that in order to determine a true resonance, the input control and output signal amplitudes must be measurements of the fundamental frequency only. Detection of resonance, and accurate measurement of the ratio of the filtered output signals to the filtered input control signal, are accomplished by filtering the input control signal and using variable low-pass or band-pass filters adjusted to eliminate all distortion of the fundamental frequency in the output circuits. In measuring filtered output signals, it is acceptable to use any meter or readout system (TRMS, average, or peak) that will accurately define the zero-to-peak amplitude of acceleration of a sine wave, since a properly filtered signal is essentially a pure sinusoid.

b. Maximum Output Acceleration. A critical frequency may be determined by monitoring the unfiltered output acceleration and noting the frequency of significant amplitude peaks. Significant peaks in the output response may occur at points other than at modes of resonance and may be more critical than resonance itself. These peak response points, if different from resonances, will occur when the signal has large amounts of distortion or if resonance is just beyond the test frequency range. Therefore, an effort should be made to define the source or cause of distortion. If the distortion is basically high-frequency hash, an effort should be made to determine if it is caused by the relative motion of parts in the test item, since resulting impacts could produce a local mechanical or functional failure.

If a single, low-order harmonic (such as second, third, fourth or fifth) is the predominant distortion on the output signal waveform, the effect on the amplitude can usually be ignored, since this would suggest a harmonic-excited resonance. The distortion would not exist if the vibration machine power supply had perfect sinusoidal fidelity and, therefore, the distorted amplitude is not a result of the test item response to the particular fundamental frequency. Also, as is the case for input signals, an output signal distorted by several harmonics simultaneously is probably the result of a nonlinear spring deflection and, most likely, a "bottoming out" nonlinear spring.

Amplitude measurements of unfiltered signals that have a large amount of high-frequency distortion tend to become "qualitative." They do not define the stress levels in a structure but, instead, indicate that something is "rattling around" with vigor, and that this particular frequency is a bad or critical frequency. If the test item has parts that are "rattling," the distortion is usually not stable, with the result that the unfiltered acceleration amplitude measurements are fluctuating and do not repeat from one cycling sweep to another or from one like test item to another. However, it is only logical to have the measuring circuits as accurate as possible and calibrated to indicate the zero-to-peak amplitude of acceleration of a sine wave. Accuracy is required in measuring the fundamental response or transmissibility of the test item. Perhaps more important than a precise acceleration amplitude measurement of highly distorted signals is a thorough investigation of the source and a frequency analysis of the distortion.

c. Monitoring Functional Operation of a Test Item. For some tests, critical frequencies can be determined by monitoring the functional operation of the item actually being tested. Examples of an indicated failure or malfunction detected by this method are chatter of switch and relay contacts, motor speed, noise level of tubes, voltage and current fluctuations, etc.

d. Miscellaneous Detection Methods. Because of the special nature of a test item detection devices other than those commonly encountered may be used to determine critical frequencies. Some examples of these devices are displacement and velocity transducers, strobe lights, high-speed motion photography, visual observation, auditory means, strain gauges, or even changes in power needed to drive the vibration machine.

Once the resonant or critical frequency has been determined, the method used for determination should be documented in the test report. If the detailed test specification requires sustained vibration at a critical frequency for a specific period of time, it is important that the original characteristics used to determine this critical frequency should be frequently monitored. If the critical frequency should shift at any time during the test, the input control frequency can be varied to maintain the new critical frequency.

COMPLEX WAVE VIBRATION TESTS

Scope

General guidance and suggested minimum requirements for conducting complex wave vibration tests are given.

A complex wave is a nonsinusoidal wave composed of several frequencies. Examples of several types of complex signals are: music, random, square, and distorted sine waves. However, this section is concerned with three specific types of inputs: random, combined sinusoidal and random, and tape recordings of actual environments or artificially generated signals.

Test Specification Check List

Certain information is required to assure a properly conducted complex wave vibration test. The points below should be listed or considered the preparation of the test specifications:

Flat Spectrum

The following items must be specified:

1. Power spectral density level (g^2/cps).
2. Frequency range.
3. Time of vibration.
4. Tolerance on equalization.
5. Identify axis (or axes) of vibration.
6. The rate (dB/octave) at which the power spectral density (PSD) outside of the desired input bandwidth is to be attenuated. (Generally, an attenuation rate no less than 18 dB/octave should be specified.)
7. Minimum proof of equalization and narrow-band analysis records required.

The following need not always be specified, but should be considered:

8. Is sequence of axes important?
9. Does each individual test item have to be equalized?
10. Output recordings.
11. Amplitude level during preliminary equalization.
12. Overall rms acceleration level.

Shaped Random

The following items must be specified:

1. Complete description, including minimum allowable slope on abrupt changes in power spectral density level.
2. Tolerance on shaping.

Random Plus Sine Tests

The following items must be specified:

1. Power spectral density level of random.
2. Peak or rms acceleration level of sine.
3. Frequency range of sine taking displacement of sine at low frequency into account.
4. Cycling time and sweep rate of sine wave.
5. If random is shaped, is sine to be flat or shaped?

Instrumentation.

Instrumentation should have the following characteristics or should be corrected (either by adjustment or corrected reading) to meet the following requirements throughout the test range:

Piezoelectric Accelerometers

1. Frequency response flat within ±5% from 5 cps to 2 Kcps.
2. Amplitude linearity of ±2%.
3. Cross-axis sensitivity of less than 5%.

True RMS Voltmeters

Used for acceleration measurements.

1. Measure the true rms value of a complex signal to a maximum crest factor (ratio of peak to rms) of 4.5.
2. Accuracy of ±3% of reading from 15 to 10,000 cps and ±5% from 5 to 15 cps.
3. The time constant of the meter circuit should be 1 sec or greater.

Cathode Followers and Amplifiers

Used in vibration amplitude measuring circuits:

1. Frequency response flat within ±2% from 10 to 10,000 cps.
2. Amplitude linearity of ±1%.

Low Pass Filters

1. Frequency response flat within ±2% between 50 and 2000 cps.
2. Frequency response down at least 2 dB at 3000 cps with a roll-off of at least 18 dB per octave.

Variable Low-Pass Filters

1. Amplitude error caused by filter should be less than ±2%.
2. Roll-off rate of at least 18 dB per octave.

Frequency Indicators

The frequency indicators used should measure the fundamental frequency within ±1.5% or ±2 cps, whichever is greater.

Random Noise Generators

Used for input vibration signal generators.

1. Uniform voltage spectral density level from 30 to 3000 cps within ±1 dB.
2. Approximately normal or Gaussian distribution of instantaneous amplitudes as verified by voltmeter readings as follows: The ratio average volts/true rms volts should lie between 0.77 and 0.83.

Fixtures

For complex wave fixture design, see to the section on sinusoidal testing.

Test Operation and Equipment

1. Equipment. The test operation and equipment used to conduct a complex wave test are different from those used to conduct a sinusoidal wave test. Primarily, this is because a complex wave vibrates the test item at several frequencies simultaneously, whereas a sinusoidal wave vibrates the test item at only one frequency at a time. In a sine wave test a flat frequency response at the table is obtained by adjusting the power level at any given frequency. In a complex wave test, the entire system from the signal source to the test item input must have a flat frequency response to reproduce the signal. There are two general types of equalization systems used to give a flat frequency response under the varying conditions of load the test item resonance. These two types are as follows:

    a. *A variable-frequency active-network equalizer or "peak-notch" equalizer.* This system consists of a number of equalizer sections. There is generally an exciter equalizer to compensate for the exciter characteristics, and several "peak-notch" equalizers to compensate for the test item characteristics. Equilization is usually checked by measuring the amplifier input voltage with a sine wave input while maintaining a low constant acceleration level at the input accelerometer. For this type of equalization, the table must be controlled on the input fundamental and, therefore, the input accelerometer signal must filtered as described in the section on sinusoidal vibration tests.

    b. *A fixed frequency type of equalization system* uses passive narrow bandwidth filters to cover the spectrum. Systems are available that split the spectrum into a variety of bandwidths. An analyzer included that has filters identical to the equalizer section so that the power spectral density level in any bandwidth can be measured. Equalization is accomplished by using a low-level, flat, random signal and adjusting at attenuator on each equalization filter so that the level is the same in all bandwidths, or at the desired level for shaped spectrums.

The requested test specification should be met within some reasonable tolerance, based on the request g^2/cps input level. Normally, a tolerance level of ± 2 dB ($+58\%$, -37%) may be considered reasonable, but an attempt should be made to keep the actual tolerance as low as practicable. The specified tolerance should be based on the results indicated by a narrow-band analysis and not based on the tolerance indicated during the equalization process.

In order to fall within a specified tolerance on a narrow-band analysis, it is necessary to hold the tolerance on equalization to a smaller magnitude. Generally, a tolerance of less than 1 dB is required on the control console analyzer so that a ± 2 dB tolerance may be approached on the narrow-band analysis plots.

Because of the statistical nature of random vibrations, an absolute tolerance applied to random amplitude measurements is not applicable. In fact, it is meaningless to specify a tolerance without specifying how the tolerance is to be determined.

The specified tolerance for verification of the equalization of the input and also for the analysis of response data shall be considered met if the narrow-band analysis plot falls within the specified tolerance (± 2 dB) when analyzed in the following manner:

Narrow-band analysis parameters

$$B - 25 \text{ cps}$$

$$50 \ BT \ 100$$

$$q - \frac{B}{4K}$$

Definition of terms

B = analyzer filter bandwidth (cps)
T = Sample length or true integration time (seconds)
K = RC time constant (seconds)
K = $\dfrac{T}{2}$
q = swept filter scan rate (cps/sec)

There may be instances where a more accurate analysis may be desired for better definition of the spectrum. In these instances, additional analysis runs that deviate from these analysis parameters are encouraged.

Equalization records should be made for review and should include the description of filter bandwidth and integration time (or RC time constant) as part of the test data to help interpret the test results. During the test, a tape recording of the input accelerometer signal should be made and the tape should be analyzed with a suitable narrow-band wave analyzer to assure that the equalization is satisfactory.

At the start of a test series on a specific test item, this check should be made for each axis. In addition, on a long test series this check should be made occasionally during test operations. Prior to the start of a new test series (that is, new fixture, new test item, new hardware, etc.), a check should be made by using the new fixture and its associated hardware (without the test item) to determine whether significant levels of energy appear in the narrow-band analysis at frequencies within and beyond the specified test frequency range. If significant levels appear, an effort should be made to define the source. Energy that can be identified as feedback from the test item is of no concern but, if the vibration system or fixture is the source of energy, it should be corrected.

2. Input Signal Sources

a. *Random Vibration.* A random vibration test is a test where the input vibration is a random noise signal. This type of signal excites several frequencies simultaneously with random amplitudes. For a stationary random signal, voltmeter indications of rms amplitude or average amplitude tend to vary about a mean value. This variance decreases as longer RC time constants are used in the voltmeter circuits. For random vibration voltmeter measurements of rms amplitude the meter time constant should be at least one sec to give a stable measurement.

The amplitude versus frequency characteristics are measured by slowly sweeping the frequency range with a narrow band-pass filter. With a flat random signal the rms amplitude is proportional to the square root of the filter bandwidth. A term has been devised to specify the amplitude that is independent of the bandwidth used to measure it. By dividing the rms amplitude by the square root of the bandwidth, the amplitude can be specified in rms $g/\sqrt{\text{cps}}$ units. Generally, this term is squared (g^2/cps) and is usually called power spectral density (PSD), although this term is only indirectly associated with power. The rms acceleration for any random wave is equal to the square root of the area under a curve of a plot of PSD (g^2/cps) versus frequency. When the PSD is flat between frequencies f_1 and f_2 (see Figure 4) spectrum is made up of a combination of the individual spectra just discussed.

The total rms acceleration in g is numerically equal to

$$G_{rms} = \sqrt{A_p + A_f + A_n}$$

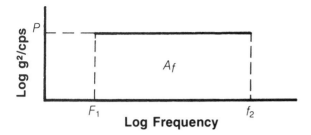

Figure 4

To set up the amplitude of random wave with the passive multiple filter equalization system, the power spectral density is set up through each filter. The overall rms acceleration, calculated from the complete spectrum, should be checked to be sure that the equalization is satisfactory. To set up the amplitude of a random wave on the active filter peak-notch systems, the overall rms acceleration is calculated and the gain is set on the amplifier to give the correct acceleration value. For all checks of the overall acceleration, the accelerometer signal should be filtered about 1.5 times the maximum test frequency to ensure that any high-frequency feedback from the test item is not reflected in the input signal measurement.

b. *Random Plus Sine Vibration.* The rms acceleration level from a sine plus random signal is calculated by squaring both rms amplitudes, adding them, and then taking the square root of the sum. For example, a 10 rms *g* sine wave mixed with a 10 rms *g* random wave would give a 14.1 rms *g* overall value. If the sine wave is specified in peak *g*, it must be converted to rms *g* before squaring and summing.

It is desirable to have an oscillator-tracking band-pass filter in the console for sinusoidal plus random wave vibration tests. If this equipment is available, the random wave amplitude is set on a meter with the noise generator, and then the sine wave amplitude is set with the oscillator on another meter through the tracking filter, with the sine wave controlled by the servo control. A correction for the random signal also included in the bandwidth of the tracking filter must be applied to the signal level controlled by the servo. The total amplitude should be checked by the rms calculation. This method gives very close control of the sine wave portion of the test which otherwise would see the same loose equalization tolerance of the random wave.

If a tracking band-pass filter is not available, the test is set up differently. The rms acceleration of each of the waves is used to calculate

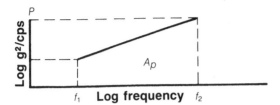

Figure 5

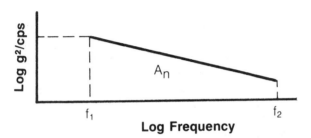

Figure 6

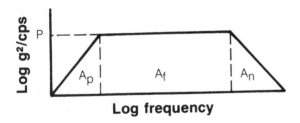

Figure 7

the rms sum. The gain on the signal generator for the wave with the lowest rms acceleration level is then set to the correct value. The gain on the other signal generator is then advanced until the rms sum is correctly set. These settings are then maintained throughout the test. For example: assume that a test requires a 0.05 g^2/cps random wave covering a frequency range from 20 to 2020 cps and 6.6 rms g sine wave is to be swept between 20 and 200 cps. The acceleration level of the random wave can be found as follows:

$$0.05 \frac{g^2}{cps} \times 2000cps = 100g^2$$

Taking the square root of this value ($\sqrt{100g^2}$) gives 10 rms g for the random wave level. The rms sum would be as follows:

$$(10\,\text{rms}\,g)^2 + (6.6\,\text{rms}\,g)^2 = 100\,\text{rms}\,g^2 + 44\,\text{rms}\,g^2 = 144\,\textit{rms}\,g^2$$

The square root of this value ($\sqrt{144\,\text{rms}\,g^2}$) gives 12 rms g for the rms sum of the sine plus random wave. This test would be set up by taking the lower of the two levels (the 6.6 rms g sine wave) and setting it up with the sine wave oscillator and then advancing the gain on the random noise generator until the overall rms is 12 g.

  c. *Tracking filters* (used for sine wave control during sine wave equalization or random plus sine tests).

(1) Bandwidth not over 25 cps at the 3 dB points at any frequency in the test range.
(2) Frequency response, including error caused by inaccurate tracking, within ±5% from 20 cps to 2 kc.

Data Interpretation

1. Acceleration Measurements. Acceleration measurements on complex wave tests tend to become meaningless unless the signal is analyzed on a wave analyzer. The method generally used for random signals is to record the signal on tape and play it into a wave analyzer. This wave analysis will show the resonant frequencies of the test item and, in general, the frequency response of the test item. The square of the response ratio at any given frequency can be obtained by comparing the input and output PSD levels.

  2. Narrow-Band Analysis. Narrow-band analysis is required to check equalization, transverse motion using random equalization, and output measurements. As the bandwidth of an analyzer is reduced, it takes a longer averaging time to get a reliable estimate of the random level. Therefore, the analysis rate, averaging time, and bandwidth should be checked to assure an accurate analysis.

STEADY-STATE ACCELERATION TESTS

General guidance and suggested minimum requirements for conducting steady-state acceleration tests within the acceleration levels from 0 to

300 g are given. Most steady-state acceleration tests are run on centri-
fuges. Some common difficulties with such testing are as follows: too
large an acceleration variation across the test item (using too small a
centrifuge), inaccurate speed control and readout, rotational instability,
and slip-ring noise which is too high. Often, a steady-state acceleration
test is run on a centrifuge too small for the job. This results in a large
acceleration variation across the test item. It is good practice to make
the variation as small as possible (less than ±10% of the specified acceleration). If the test item functions as a result of applying acceleration,
the acceleration variation may need to be held much smaller than
±10%. Also, the orientation of the test item on the centrifuge may be
critical. With such devices, the point at which the acceleration is to be
applied must be specified. In many instances, the test should be
recorded with an accelerometer and suitable readout instrumentation in
order to have an operational check on the centrifuge speed control and
readout system.

Test Specification Check List

Certain information is required to assure a properly conducted steady-
state acceleration test. The following points should be listed or considered in the preparation of the test specifications:

1. The level of acceleration to be applied and its tolerance (an absolute
   tolerance of ±5% is reasonable).
2. Direction of applied acceleration[2] (a sketch is desirable).
3. Minimum (or maximum if applicable) duration of applied acceleration (an absolute tolerance of ±10% of the specified duration is
   reasonable).
4. Sequence of tests.
5. Under what conditions and intervals the functional tests (specify
   what) are to be performed.
   The following items need not always be specified but should be considered:
6. Is the rate of change of acceleration important? (Do not be more
   restrictive than necessary as it is usually much more difficult to control the rate of change than the final acceleration.)
7. If the acceleration level at any specified internal point of the test
   item is critical, that point should be located with reference to
   known external points.
8. Is the normal allowable ±10% variation across the test item with
   respect to the specified acceleration suitable? (On a large item this
   may be impractical. On acceleration-sensitive devices, a smaller
   variation may be required.)

9. Are cables or attachments to be tied down in some specified manner?
10. Is there anything unique about the mounting?

Instrumentation

The most stringent requirement for an acceleration test is an accurate measurement of the rotational velocity of the centrifuge arm or platform indicated in rpm.

Two types of instrumentation are generally used; the tachometer and the magnetic pickup. Either system should have an accuracy of no less than ±1%. This accuracy may be confirmed by an accelerometer and readout system. The following equation may be used for computation:

$$\text{rpm} = 187.76\sqrt{g/r}$$

where

rpm = calculated rotational speed of the centrifuge
g = measurement of acceleration
r = distance of accelerometer mounting from center of rotation (inches).

The preferred rpm indication system consists of a spur gear attached to the centrifuge spindle, a magnetic pickup, and an electronic counter. The gear should have at least 600 spurs or teeth. Mounting of the magnetic pickup should be as close as practicable to the gear and be rigid to minimize shifting or displacement. A decade counter provides direct readout if the number of gear teeth is 6, 600, 6000, etc. The counter should be accurate to ±1 count. The electronic counter may be calibrated by supplying a frequency signal accurate to ±1% to input of the counter.

If direct acceleration measurements are made, the accelerometer must be accurately calibrated to a tolerance of ±0.5% or better. Other disadvantages of this system are: (a) the location of the sensitive element of the accelerometer must be known exactly, (b) the radial distance of the accelerometer mounted on the centrifuge must be known accurately, and (c) the noise level in the readout system must be less than 0.1% of the accelerometer signal. The most probable source of noise is the slip-ring system.

Another possible method to measure rotational velocity is the use of a stroboscopic light source. The difficulty and hazards of observing the rotating arm make this undesirable. In addition, below 300 rpm it is subject to large errors by operators. For example, an uncertainty of one rpm at 100 rpm and a radius of 36 in. gives an uncertainty of 0.2 of a g acceleration which is 2% of a nominal 10 g.

Fixtures

The requirements for fixtures are strength and lightness. The materials and construction of mounting fixtures should follow good design practices to avoid destruction under the maximum stresses to be imposed. They should be light enough so that the combined weights of fixtures, components, and mounting fasteners do not exceed the capabilities of the centrifuge.

Test Operations and Equipment

The test unit is to be attached to the centrifuge at the distance from the center of rotation of centrifuge arm or table that will provide the specified acceleration. The following equations may be used to compute either rpm or the point of mounting:

$$\text{acceleration } (g) = 0.0000284\, r\ (\text{rpm})^2 \text{ or}$$

$$r = g \div 0.0000284\ (\text{rpm})^2 \tag{1}$$

$$\text{rpm} = 187.76\sqrt{g/r} \tag{2}$$

where

rpm = revolutions per minute
g = acceleration level
r = distance in inches from center of rotation of centrifuge arm or table

    The acceleration variation across the test item should be computed by measuring the points of the item nearest to and farthest from the center of the centrifuge, substituting these values in Eq. (1) and computing the acceleration levels for these points.

    Constant rotational speed and accurate indication of the speed are imperative. Some sample computations using the aforementioned equations will show that a $\pm1\%$ variation in speed results in approximately $\pm2\%$ change in acceleration level. If the speed indication has only $\pm1\%$ accuracy, then the speed control and stability must be better than $\pm1.5\%$ in order to remain within a test tolerance of $\pm5\%$.

    It is recommended that the instrumentation used to measure acceleration have an accuracy of $\pm2\%$. To meet this tolerance, the noise level of the centrifuge slip-ring must be kept at a very low level. This requires maintenance and care in cleaning and lubricating slip-rings and contacts. Slip-ring noise levels should be periodically monitored by an oscilloscope or equivalent measuring equipment.

Data Interpretation
Revolutions per minute, the acceleration level, and the mounting distance, or radius, of the test item must be recorded.

SUMMARY
We have discussed the philosophy necessary for a successful environmental test program. Detailed specification, required instrumentation, test methods (with examples), data collection and its interpretation have been given. The environments covered were vibration and acceleration.

RECOMMENDED REFERENCES

Sinusoidal Vibration Tests

Roark, R.J. *Formulas for Stress and Strain.* New York: McGraw-Hill, 1954. This volume contains information on stiffness of members or structures.

Harris, C.M. and Creed, C.E. *Shock and Vibration Handbook.* Vols. I-III. New York: McGraw-Hill, 1961. This handbook contains general information concerning the art of vibration testing.

U. S. Atomic Energy Commission. *ALO Standardization Program Handbook.* AL Appendix 72XF. Albuquerque Operations Office. June 1, 1961.

Complex Wave Vibration Tests

Morrow, Charles T. Averaging Time and Data Reduction Time for Random Vibration Spectra. *Journal of the Acoustical Society.* June 1958.

Miles, John W. and Thompson, W.T. Statistical Concepts in Vibration. *Shock and Vibration Handbook,* Harris and Creed, New York; McGraw-Hill, 1961.

Harris, Cyril M. Introduction and Terminology *Handbook of Noise Control,* Ch. 1, New York: McGraw-Hill, 1957.

Hillyer, D.F., Jr. Random Vibration Terminology and Analysis Equipment. Sandia Corporation Monograph SCL-R-64-90. February 1964.

U.S. Atomic Energy Commission. *ALO Standardization Program Handbook.* AL Appendix 72XF. Albuquerque Operations Office. June 1, 1961.

NOTES

[1]This equation is not applicable when $R = 3$; for $R = 3$, area = $A_2 = 23 - f_i \log (f_2/f_1)$ or $A_2 = Pf_1 \ln(f_2/f_1)$.

[2]The direction of applied acceleration is inward toward the center of rotation.

Chapter 9

Sunshine, Sand and Dust, Explosion, and Salt Fog Environments

In Chapters 7 and 8 we discussed the "mechanical type" of environments. Now we discuss the "weather and exposure types" that we feel will be most deleterious to a product and, even though there are many, we have chosen the ones that we feel the product will most frequently encounter and that should be considered during the initial design of the product. An excellent military specification on depicting environmental test methods is MIL-STD-810. We have reformulated portions of this specification in order to present a concise, easy-to-follow methodology for sunshine, sand and dust, rain, explosive atmosphere, and salt spray.

SUNSHINE

The sunshine test is conducted to determine the effect of solar radiation energy on equipment in the earth's atmosphere. For the purpose of this test, only the terrestrial portion of the solar spectrum is considered. The specified limits and energy levels provide the simulated effects of natural sunshine. The ultraviolet portion simulates natural sunshine in a general way, and is considered representative of irradiation in most geographical locations. Sunshine causes heating of equipment and photo-degradation, such as fading of fabric colors, checking of paints, and deterioration of natural rubber and plastics. Sunshine tests are applicable to equipment that may be exposed to solar radiation during service or unsheltered storage at the earth's surface or in the lower atmosphere.

Equipment

Test chamber volume must be a minimum of ten times that of the volume of the envelope volume of the test item. The chamber's simulated solar radiation source area must be a minimum of 125% of the horizontal area projection of the test item.

Solar Radiation Source

For the purposes of this test, the following spectral distribution of solar radiation is acceptable: 50 to 72 watts/ft^2 of infrared (of wavelengths above 7800 angstrom units), 4 to 7 watts/ft^2 of ultraviolet (of wavelengths below 3800 angstrom units), and the balance visible. Radiation sources must be located at least 30 inches away from any outer surface of the test item. Lamp vendor's spectral distribution curves may be used in establishing the spectral distribution.

Tests conducted for degradation and deterioration of materials and for heat build-up within the test item may use one of the following acceptable radiation sources:

1. Mercury vapor lamps (internal reflector type only).
2. Combination of incandescent spot lamps (including infrared filters) together with tubular type mercury vapor lamps with external reflector.
3. Combination of incandescent spot lamps (including infrared filters) together with mercury vapor lamps (with internal reflectors with filters are required).
4. Carbon arc lamps with suitable reflectors.
5. Mercury-xenon arc lamps with suitable reflectors (with filters are required).

Tests that are conducted only for heat build-up within the test item may use infrared lamps of the incandescent type or other radiant heating source approved by the procuring activity.

TEST METHODS

Method One. Accelerated Steady-State, Solar Radiation Test Nonoperating

The test item must be placed in the test chamber and exposed to radiant energy at the rate of 100 to 120 watts per square foot or as specified in the equipment specification. The period of the test must not be less than the 48 hours, during which time the chamber temperature must be maintained at 45°C ±3°C (113°F). At the conclusion of the exposure period, and with the chamber temperature maintained as specified, the

test item must be operated and the results compared with data previously obtained. The test item must then be returned to room temperature and inspected.

**Method Two. Simulated Cycling Temperature and Solar Radiation Test,
Operating**

Test items must be placed in the test chamber and a performance pretest
must be conducted. Test items must then be exposed to five continuous
24-hour cycles of controlled simulated solar radiation and dry-bulb temperature (as indicated in Figure 1) or as specified in the equipment specification. Tolerances for control of radiation must be ±10 watts/ft².
Tolerances for air temperature control must be ±3°C. Standard ambient conditions must be maintained within the test chamber at the beginning and end of each test cycle except that the chamber relative humidity (uncontrolled) must be a maximum of 40% (equivalent to 50 grains
moisture/pound dry air) at standard ambient temperature. The air velocity in the chamber must be maintained within 3 to 6 knots (300 to

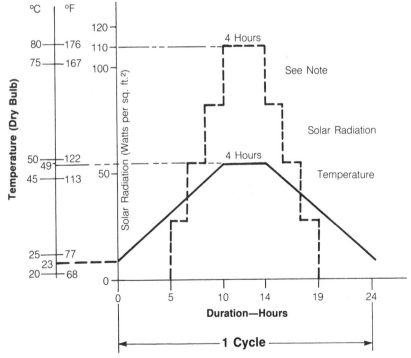

Figure 1 Simulated solar radiation cycle (method two). [*Note*: Use a minimum of three intermediate equal steps for increase or decrease of simulated solar
radiation intensity.]

600 ft/min).   At the conclusion of the last exposure, the test items must be removed from the chamber and operated and the results compared with the data previously obtained during pretest.   The test items must then be returned to room ambient and inspected.   (*Note:* Test items may or may not be operated during the test at the option of the quipment specification.)

SAND AND DUST

Sand and Dust is a test used during the development and evaluation of equipment to ascertain its ability to resist the effects of a dry (fine sand) laden atmosphere.   This test simulates the effect of sharp-edged dust (fine sand) particles, up to 150 microns in size, which may penetrate into cracks, crevices, bearings, and joints, and cause a variety of damage such as fouling moving parts, making relays inoperative, forming electrically conductive bridges with resulting "shorts" and acting as a nucleus for the collection of water vapor, and hence a source of possible corrosion and malfunction of equipment.   This test is applicable to all mechanical, electrical, electronic, electrochemical, and electromechanical devices for which exposure to the effects of a dry dust (fine sand) laden atmosphere is anticipated.

Equipment

The test facility must consist of a chamber and accessories to control dust concentration, velocity, temperature, and humidity of dust-laden air.   In order to provide adequate circulation of the dust-laden air, no more than 50% of the cross-sectional area (normal to air flow) and 30% of the volume of the chamber must be occupied by the test item(s). The chamber should be provided with a suitable means of maintaining and verifying the dust concentration in circulation.   A minimum acceptable means for doing this is by use of a properly calibrated smoke meter and standard light source.   The dust-laden air must be introduced into the test space in such a manner as to allow it to become approximately laminar in flow before it strikes the test item.

   Dust used in this test must be a fine sand (97 to 99% by weight SiO_2) of angular structure, and must have the following size distribution as determined by weight, using the U.S. Standard Sieve Series: 100% of this dust should pass through a 100-mesh screen; 98 ± 2% of the dust should pass through a 140-mesh screen; 90 ± 2% of the dust should pass through a 200-mesh screen; and 75 ± 2% of the dust should pass through a 325-mesh screen.

   In the performance of these tests, "140-mesh silica flour," as produced by the Fenton Foundry Supply Company, Dayton, Ohio, and Ottawa Silica Company, Ottawa, Illinois, or equal, is satisfactory.

Method

Place the test item in the chamber, positioned as near the center of the chamber as practicable. If more than one item is being tested, there must be a minimum clearance of four inches between surfaces of test items or any other material or object capable of furnishing protection. Also, no surface of the test item must be closer than four inches from any wall of the test chamber. Orient the item so as to expose the most critical or vulnerable parts to the dust stream. The test item orientation may be changed during the test if so required by the equipment specification.

1. Set the chamber controls to maintain an internal chamber temperature of 23°C (73°F) and a relative humidity of less than 22%. Adjust the air velocity to 1750 ±250 feet per minute. Adjust the dust feeder to control the dust concentration at 0.3 ±0.2 g per cubic foot. With the test item nonoperating, maintain these conditions for six hours.

2. Stop the dust feed and reduce the air velocity to 300 ±200 feet per minute. Raise the internal chamber air temperature to 63°C (145°F) and adjust humidity control to maintain a relative humidity of less than 10% . Hold these conditions overnight (approximately 16 hours).

3. While holding chamber temperature at 63°C (145°F), adjust the air velocity to 1750 ±250 feet per minute. Adjust the dust feeder to control the dust concentration at 0.3 ±0.2 g per cubic foot. With the test item nonoperating, maintain these conditions for six hours.

4. Turn off all chamber controls and allow the test item to return to standard ambient conditions. Remove accumulated dust from the test item by brushing, wiping, or shaking, care being taken to avoid introduction of additional dust into the test item. Under no circumstances should dust be removed by either air blast or vacuum cleaning.

5. Operate the test item and compare the results with data previously obtained.

6. Inspect the test item. In the performance of this inspection, test items containing bearings, grease seals, lubricants, etc., must be carefully examined for the presence of dust deposits.

[*Note:* The test specimen may be operating during either or both of the six-hour test periods (step 1 or 3) if so required by the equipment specification.]

RAIN

The rain test is conducted to determine the effectiveness of protective covers or cases to shield equipment from rain. This test is applicable to equipment that may be exposed to rain under service conditions.

Where a requirement exists for determining the effects of rain erosion on radomes, nose cones, etc., a rocket sled test facility or other such facility should be considered. Since any test procedure involved would be contingent on requirements peculiar to the test item and the facility employed, a standardized test procedure for rain erosion is not included in this test method.

Equipment

Rain chambers must have the capability of producing falling rain and a facility for producing wind blowing at the rates specified herein. Chamber temperature must be uncontrolled, except as regulated by water introduced as rain, throughout the test period. Rain should be produced by a water distribution device of such design that the water is emitted in the form of droplets having a diameter range between 1 and 4 mm. The temperature of the water should be between 11 and 35°C (52 and 95°F). Water distribution must be such that, with the wind source turned off, the rain is dispersed over the test area within the limits specified below. The wind source should be so positioned, with respect to the equipment, that it will cause the rain to beat directly and uniformly against one side of the equipment. The wind source must be capable of producing horizontal wind velocities up to 40 miles per hour. Wind velocity should be measured at the position of the test item, prior to placement of the item in the chamber. No rust or corrosive contaminants must be imposed on the test item by the test facility.

[The recommended method of measuring raindrop size is the flour pellet method as indicated in "The Relation of Raindrop Size to Intensity" by Laws and Parsons, Transactions of the American Geophysics Union, Part II, pages 452-459 (1943).]

Method

Test items must be placed in the chamber in its normal operating position. Test items should be exposed to a simulated rain at a rate of 5 ±1 inches per hour for 10 minutes. The rate of rainfall shall then be raised to 12 ±1 inches per hour and held at this rate for 5 minutes. The rate must then be reduced to 5 ±1 inches per hour for the next 15 minutes. Starting 5 minutes after the initiation of the rain, the wind source shall be turned on and adjusted to produce a horizontal wind velocity of 40 miles per hour (3500 feet per minute). Then the wind source must be turned off. [Note: If specified in the equipment specification, the equipment should be operated during the last 10 minutes of the 30 minutes rain. Each of the sides of the test items that could be exposed to blown rain must be subjected to the rain for a period of not less than 30 minutes, for a total test duration of not less than 2 hours.] At the conclusion of the test period, the test items must be removed from the test chamber, operated, and the results compared with those previously obtained. Protective covers or cases should, where possible, then be removed and the items inspected, with particular attention to evidence of water penetration, such as free water, swelling of material, or other deterioration.

EXPLOSIVE ATMOSPHERE

Explosive atmosphere testing is conducted to determine the ability of equipment to operate in the presence of an explosive atmosphere without creating an explosion or to contain an explosion occurring inside the equipment. Since equipments operate in ever-changing potentially explosive atmospheres, the equipments, when being laboratory tested, must operate in the presence of the optimum fuel-air mixture which requires the least amount of energy for ignition. Equipment igniting energy may be produced electrically, thermally, or chemically.

Equipment

An explosion-proof test chamber equal to that specified in MIL-C-9435 should be used. Unless otherwise specified, the fuel must be gasoline, grade 100/130 conforming to MIL-G-5572. The equipment used to vaporize the fuel for use in the explosion-proof test should be so designed that a small quantity of air and fuel vapor will be heated together to a temperature such that the fuel vapor will not condense as it is drawn from the vaporizer into the chamber.

Sample Vapor Calculation

An illustration of the procedure for calculating the weight of 100/130 octane gasoline required to produce the desired 13-to-1 air-vapor ratio, the following sample problem is presented. This information is required:

1. Chamber air temperature during test: 27°C (80°F).
2. Fuel temperature: 24°C (75°F).
3. Specific gravity of fuel at 16°C (60°F): 0.704.
4. Test altitude: 20,000 feet (P = 6.75 lb/in.^{2}).
5. Air-vapor ratio (desired): 13-to-1.

Step 1

Employing the following equation, calculate the apparent air-vapor ratio:

$$AAV = \frac{AV_{(desired)}}{1.04(\dfrac{P}{14.696}) - 0.04} = \frac{13}{1.04(\dfrac{6.75}{14.696}) - 0.04} = 29.68$$

where

P   = pressure equivalent of altitude, lb/in.^{2}
AAV = apparent air-vapor ratio
AV   = desired air-vapor ratio

At or above 10,000 feet altitude, with chamber air temperature above 16°C (61°F) and at AV ratio of five or greater, air-vapor ratio = air fuel ratio (AF) for 100/130 octane fuel. Since the conditions of the explosion test under consideration will always be well above these values, AV will equal AF in all cases.

Step 2

Since AV = AF, use Figure 2 to determine the weight of air (WA) and divide by AAV to obtain uncorrected weight of fuel required (W_{fu}).

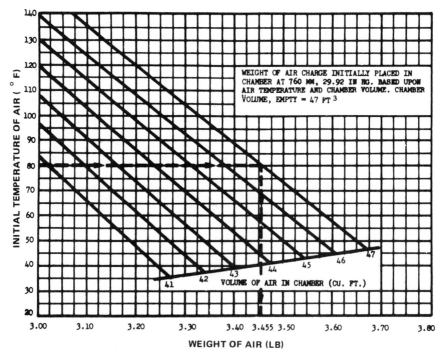

Figure 2   Weight of air charge versus temperature.

$$W_{fu} = \frac{WA}{29.68} = \frac{3.455}{29.68} = 0.116 \text{ lb, fuel weight (uncorrected)}$$

Figure 2 pertains to a specific test chamber and should not be used for all test facilities. It is utilized here for illustration of the method of employment only. Each test chamber must have its own chamber volume chart.

Step 3

Knowing fuel temperatures and specific gravity at 16°C (61°F), use Figure 3 to determine specific gravity at given temperature.

Step 4

Using Figure 4, obtain correction factor K for the specific gravity determined during step three. Apply factor to obtain weight of fuel corrected (W_{fc}).

$$W_{fc} = KW_{fu} = 1.01 \times 0.116 = 0.117 \text{ lb, fuel weight (corrected).}$$

TEST METHODS

Method One

This method is intended for determining the explosion producing characteristics of equipment not hermetically sealed and not contained in cases designed to prevent flame and explosion propagation. Ground equipment used in or near vehicles must also be tested in accordance with this procedure, except that the specified altitude survey need be conducted only to 10,000 feet.

1. The test item must be installed in the test chamber and in such a manner that normal electrical operation is possible and mechanical controls may be operated through the pressure seals from the exterior of the chamber. External covers of the test item should be removed or loosened to facilitate the penetration of the explosive mixture. Large test items may be tested one or more units at a time, by extending electrical connections through the cable port to the balance of the associated equipment located externally.
2. The test item must be operated to determine that it is functioning properly and to observe the location of any sparking or high temperature components that may constitute potential explosion hazards.
3. Mechanical loads on drive assemblies and servomechanical and electrical loads on switches and relays may be simulated when neces-

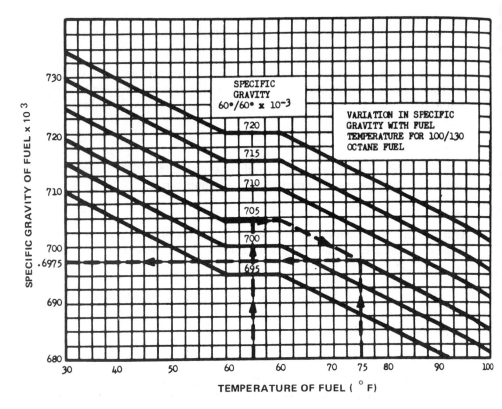

Figure 3 Specific gravity versus temperature.

sary if proper precaution is given to duplicating the normal load in respect to torque, voltage, current, inductive reactance, etc. In all instances, it is considered preferable to operate the test item as it normally functions in the system during service use.

Test Procedure I

The test must be conducted as follows at simulated test altitudes of ground level to 5000 feet, 10,000 feet (10,000 feet maximum for ground equipment), 20,000 feet, 30,000 feet, 40,000 feet, and 50,000 feet. (Pressures for altitudes are given in the U.S. Standard Atmosphere, 1962.)

Step 1

Test chambers must be sealed and the ambient temperature within must be raised to 71°C ±3°C (160°F), or to the maximum temperature to which the test item is designed to operate (if lower than 71°C or 160°F).

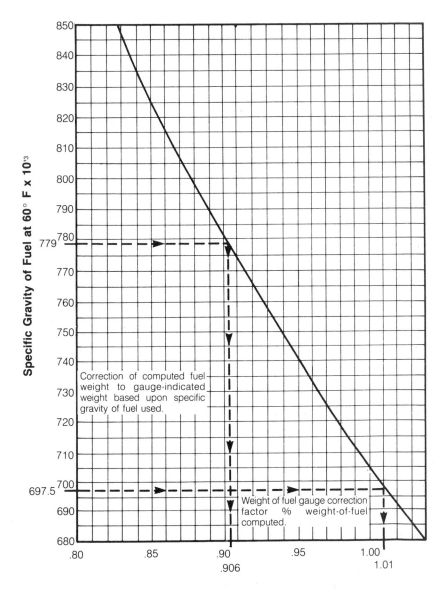

Figure 4 Fuel weight to gage indicated weight correction factor.

Temperature of the test item and the chamber walls must be permitted to rise to within 11°C (20°F) of that of the chamber ambient air, prior to introduction of the explosive mixture.

Step 2

Internal test chamber pressure must be reduced sufficiently to simulate an altitude approximately 10,000 feet above the desired test altitude. The weight of fuel necessary to produce an air-vapor ratio of 13-to-1 at the desired test altitude should be determined from consideration of chamber volume, fuel temperature and specific gravity, chamber air and wall temperature, test altitude, etc. A time of 3 ±1 minutes should be allowed for introduction and vaporization of the fuel. Air must be admitted into the chamber until a simulated altitude of 5000 feet above the test altitude is attained.

Step 3

Operation of the test item must, at this time, be commenced, all making and breaking electrical contacts being actuated. If high temperature components are present, a warm-up time of 15 minutes should be permitted. If no explosion results, air should be admitted into the chamber so as to steadily reduce the altitude down past the desired test altitude to an elevation 500 feet below that altitude but not to exceed a pressure of one atmosphere. The operation of the test item must be continuous throughout this period of altitude reduction, and all making and breaking electrical contacts must be operated as frequently as deemed practicable.

Step 4

If, by the time the simulated altitude has been reduced to 5000 feet below the test altitude, no explosion has occurred as a result of operation of the test item, the potential explosiveness of the air-vapor mixture must be verified by igniting a sample of the mixture with a spark gap or glow plug. At pressure altitudes of 20,000 feet, or higher, the attainment of ignition at any altitude must be sufficient evidence that the mixture was ignitable even though ignition was not obtained at some other point in the vicinity of the test altitude.

At any altitude below 20,000 feet, the mixture sample must ignite immediately at the point within 3000 feet of the test altitude. If the air-vapor mixture is not found to be explosive, the test must be considered void and the entire procedure repeated.

Failure Criteria

If the item causes explosion at any of the test altitudes, it must be considered to have failed the test and no further trials need be attempted.

Method Two

This method is intended for determining the explosion and flame arresting characteristics of equipment cases or item enclosures designed for that purpose.

1. When necessary, the test item case or item enclosures must be prepared for explosion-proof testing by drilling and tapping openings in the case or enclosure for inlet and outlet hose connections to the fuel vapor air mixture circulation system and for mounting a spark gap device. The case volume should not be altered by more than ±5% by any modification to facilitate the introduction of explosive vapor.
2. When inserting a hose from a blower, adequate precaution should be taken to prevent ignition of the ambient mixture by backfire or the release of pressure through the supply hose.
3. A spark gap device for igniting the explosive mixture within the case or enclosure must be provided. The case or enclosure may be drilled and tapped for the spark gap device or the spark gap device may be mounted internally.
4. The case or enclosure with either the test item or a model of the test item of the same volume and configuration in position within the case or enclosure must be installed in the explosion chamber.

Test Procedure II

The rest must be accomplished three times at altitudes between ground level and 5000 feet as follows:

Step 1

Chambers must be sealed and the internal pressure reduced sufficiently to simulate an altitude between ground level and 5000 feet. The ambient chamber temperature should be at least 25°C (77°F). An explosive mixture within the chamber must be obtained by following the procedure set forth in Procedure I.

Step 2

The internal case ignition source must be energized in order to cause an explosion within the case. The occurrence of an explosion within the case may be detected by use of a thermocouple inserted in the case and connected to a sensitive galvanometer outside the test chamber. If ignition of the mixture within the case does not occur immediately, the test must be considered void and must be repeated with a new explosive charge.

Step 3

At least five internal case explosions must be accomplished at the test altitude selected. If the case tested is small (not in excess of one-fiftieth of the test chamber volume) and if the reaction within the case upon ignition is of an explosive nature without continued burning of the mixture as it circulates into the case, more than one internal case explosion (but not more than five) may be produced without recharging the entire chamber. Ample time should be allowed between internal case explosions for replacement of burnt gases with fresh explosive mixture, within the case. If the internal case explosions produced did not cause a main chamber explosion, the explosiveness of the fuel-air mixture in the main chamber must be verified. If the air-vapor mixture in the main chamber is not found to be explosive, the test must be considered void and the entire procedure repeated.

Failure Criteria

If the internal case explosion causes a main chamber explosion, the test item must be considered to have failed the test and no further trials need be attempted.

SALT FOG

The salt fog test is conducted to determine the resistance of equipment to the effects of a salt atmosphere. Damage to be expected from exposure to salt fog is primarily corrosion of metals, although in some instances salt deposits may result in clogging or binding of moving parts. In order to accelerate this test and thereby reduce testing time, the specified concentration of moisture and salt is greater than is found in service. The test is applicable to any equipment exposed to salt fog conditions in service.

This test should be applied only after full recognition of the following deficiencies and limitations:

1. Successful withstanding of this test does not guarantee that the test item will prove satisfactory under all corrosive conditions.
2. Salt fog used in this test does not truly duplicate the effects of a marine atmosphere.

3. It is highly doubtful that a direct relationship exists between salt fog corrosion and corrosion due to other media.

4. This test is generally unreliable for comparing the corrosion resistance of different metals or coating combination, or for predicting their comparative service life. (Some idea of the service life of different life samples of the same, or closely related metals, or of protective coating-base metal combinations exposed to marine or seacoast locations can be gained by this test provided that the correlation of field service test data with laboratory tests shows that such a relationship does exist, as in the case of aluminum alloys. Such correlation tests are also necessary to show the degree of acceleration, if any, produced by the laboratory test.)

5. The salt fog test is acceptable for evaluating the uniformity (that is, thickness and degree of porosity) of protective coatings, metallic and nonmetallic of different lots of the same product, once some standard level of performance has been established. (When used to check the porosity of metallic coatings, the test is more dependable when applied to coatings that are cathodic rather than anodic toward the basic metal).

6. This test can also be used to detect the presence of free iron contaminating the surface of another metal by inspection of the corrosion products.

Equipment

Equipment used in the salt fog test should include the following: Exposure chamber with racks for supporting test items, salt solution reservoir with means for maintaining a constant level of solution, means for atomizing salt solution, including suitable nozzles and compressed air supply, chamber-heating means and control, and means for humidifying the air at a temperature above the chamber temperature.

Chambers and all accessories must be made of material that will not affect the corrosiveness of the fog such as glass, hard rubber, plastic, or kiln dried wood, other than plywood. In addition, all parts that come in contact with test items must be of materials that will not cause electrolytic corrosion. The chamber and accessories should be constructed and arranged so that there is no direct impingement of the fog or dripping of the condensate on the test items, so that the fog circulates freely about all test items to the same degree, and so that no liquid which has come in contact with the test items returns to the salt-solution reservoir.

Chambers must be properly vented to prevent pressure build-up and allow uniform distribution of salt fog. The discharge end of the vent should be protected from strong drafts that can create strong air currents in the test chamber.

Atomizers used must be of such design and construction as to produce a finely divided, wet, dense fog. Atomizing nozzles must be made of material that is nonreactive to the salt solution.

Compressed air entering the atomizers must be essentially free from all impurities, such as oil and dirt. Means should be provided to humidify and warm the compressed air as required to meet the operating conditions. The air pressure must be suitable to produce a finely divided dense fog with the atomizer or atomizers used. To insure against clogging the atomizers by salt deposition, the air should have a relative humidity of at least 85% at the point of release from the nozzle. A satisfactory method is to pass the air in very fine bubbles through a tower containing heated water which should be automatically maintained at a constant level. The temperature of the water should be at least 35°C (95°F). The permissible water temperature increases with increasing volume of air and with decreasing heat insulation of the chamber and the chamber's surroundings. However, the temperature should not exceed a value above which an excess of moisture is introduced into the chamber [for example, 43°C (109°F) at an air pressure of 12 psi] or a value that makes it impossible to meet the requirement for operating temperature.

Preparation of Salt Solution

Salt used must be sodium chloride containing, on the dry basis, not more than 0.1% sodium iodide and not more than 0.2% of total impurities. Unless otherwise specified, a 5 ±0.1% solution should be prepared by dissolving 5 ±0.1 parts by weight of salt in 95 parts by weight of distilled water. The solution must be adjusted to and maintained at a specific gravity of from 1.023 to 1.037. In order to determine whether the percentage of sodium chloride in the solution falls within the specified range, see Figure 5, utilizing the measured temperature and density of the salt solution.

The pH of the salt solution must be so maintained that the solution atomized at 35°C − 2°C + 1°C (95°F − 4°F + 2°F) and collected by the method specified in this test method will be in the pH range of 6.5 to 7.2. Only diluted C. P. hydrochloric acid or C. P. sodium hydroxide should be used to adjust the pH. The pH measurement must be made electrometrically by using a glass electrode with a saturated potassium chloride bridge or by a colorimetric method such as bromothymol blue, provided that the results are equivalent to those obtained with the

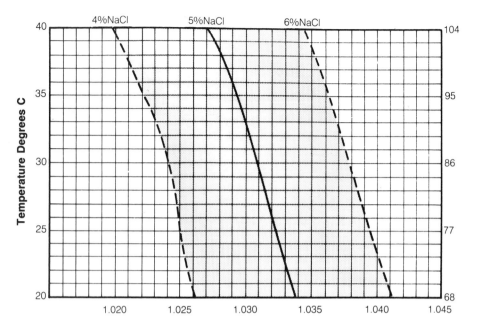

Figure 5 Variations of specific gravity of salt (NaCl) solution with temperature.

electrometric method. The pH must be measured when preparing each new batch of solution.

A filter fabricated of noncorrosive materials similar to that shown in Figure 6 should be provided in the supply line and immersed in the salt solution reservoir in a manner such as that illustrated in Figure 7.

Test Chamber Operating Conditions

Testing should be conducted with a temperature in the exposure zone maintained 35°C − 2°C + 1°C (95°F − 4°F + 2°F). Satisfactory methods for controlling the temperature accurately are by housing the apparatus in a properly controlled constant temperature room, by thoroughly insulating the apparatus and preheating the air to the proper temperature prior to atomization, or by jacketing the apparatus and controlling the temperature of the water or of the air used in the jacket. The use of immersion heaters within the chamber, for the purpose of maintaining the temperature within the exposure zone, is prohibited.

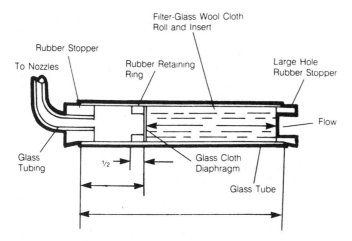

Figure 6 Salt solution filter.

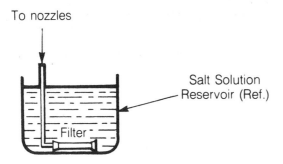

Figure 7 Location of salt solution filter.

Suitable atomization has been obtained in chambers having a volume of less than 12 cubic feet with the following conditions:

1. Nozzle pressure must be as low as practicable to produce fog at the required rate.
2. Orifices between 0.02 and 0.03 inches in diameter.
3. Atomization of approximately three quarts of salt solution per 10 cubic feet of chamber volume per 24 hours.

When using large size chambers having a volume considerably in excess of 12 cubic feet, the conditions specified may require modification to meet the requirements for operating conditions.

Salt fog conditions maintained in all parts of the exposure zone must be such that a clean fog collecting receptacle placed at any point in the exposure zone will collect from 0.5 to 3 ml of solution per hour for each 80 cm^2 of horizontal collecting area (10 cm diameter) based on an average test of at least 16 hours. A minimum of two receptacles must be used—one placed nearest to any nozzle and one farthest from all nozzles. Receptacles must be placed so that they are not shielded by test items and so that no drops of solution from test items or other sources will be collected.

Solution, collected in the manner specified in this test method, must have the sodium chloride content and pH specified in this test method when measured at a temperature of 35°C − 2°C + 1°C (95°F − 4°F + 2°F).

Salt solution from all collection receptacles used can be combined to provide that quantity required for the measurements specified. The solution, maintained at the specified temperature, can be measured in a graduate of approximately 2.5 cm inside diameter. A small laboratory type hydrometer will be required for measurement within this volume. The pH must be measured, as specified under the heading "Preparation of Salt Solution."

Measurement of both sodium chloride and pH must be made at the following specified times:

1. For salt fog chambers in continuous use the measurements should be made following each test.
2. For salt fog chambers that are used infrequently a 24-hour test run should be accomplished, followed by the measurements. The test item must not be exposed to this test run.

Preparation of Test Item

Test items must be given a minimum of handling, particularly on the significant surfaces, and must be prepared for test immediately before exposure. Unless otherwise specified, coated or uncoated metallic devices must be thoroughly cleaned of oil, dirt, and grease, as necessary, until the surface is free from water break. Cleaning methods should not include the use of corrosive solvents nor solvents that deposit either corrosive or protective films, nor the use of abrasives other than a paste of pure magnesium oxide. Test items that have an organic coating should not be solvent-cleaned. Those portions of test items which come in contact with the support and unless otherwise specified, i.e., the case of coated devices of samples, cut edges and surfaces not required to be coated, must be protected with a suitable coating of wax or similar substance impervious to moisture.

Test Procedure

Test items must be placed in the test chamber exposed to the salt fog for a period of not less than 48 hours. At the end of the test period the test items should be operated and the results compared with the data previously obtained. Salt deposits resulting from the test may be removed by such methods specified in the individual equipment specification prior to operation of the test item. The test items must then be inspected for deleterious effects. If necessary to aid in examination, a gentle wash in running water not warmer than 38°C (100°F) may be used. The test items must then be stored in an ambient atmosphere for 48 hours, or as specified in the equipment specification for drying. At the end of the drying period, the test items must again be operated and the results compared.

SUMMARY

We have discussed the philosophy necessary for a successful environmental test program. Detailed specifications, required instrumentation, test methods (with examples), and data collection and its interpretation have been given. The environments covered were, sunshine, sand and dust, explosion, rain, and salt fog.

Chapter 10

Statistical Techniques
for Product Evaluation Efficiency

The proper use of modern statistical techniques provides not only a precise summary of the conclusions that may be drawn from an evaluation already performed, but also a reliable prediction of the information that can be gained from a proposed experiment.

This chapter is aimed at readers who would like to use statistical techniques to improve the efficiency of their product evaluation experiments. Test planning, in the modern statistical sense, compels the experimenter to formulate his objectives and procedures which often leads to the recognition of fallacies and other difficulties in advance of data collecting.

The techniques described are the most efficient that are known. They are designed to maximize the amount of information obtainable from a given sample size. In addition, these techniques are sufficiently definitive to serve as standard procedures. Their uniform application is as important as their efficiency. A large part of the value of experimentally determined data is the scope of applicability. Data collected by means of standardized procedures are cumulative in the mathematical sense. Hence the precision with which performance characteristics are known can be improved with time as additional data are collected. For those readers not thoroughly familiar with statistical terms, a glossary of selected terms appears at the end of this book.

STATISTICAL CONCEPTS

Modern statistical concepts, such as probability, experimental error, population-sample relation, frequency distributions, confidence intervals, sample size, and design of experiment techniques must be understood, if economical, efficient tests and experiments are to be conducted. The vexing problem of confidently demonstrating component and product performance with small sizes can be solved only with these concepts.

EXPERIMENTAL ERRORS

Contemporary statistical methods of testing contain an ingredient, not explicit in mathematics, called *error*. This philosophy assumes that there is an error in every measurement made, and, as a consequence the true values of measurable characteristics can never be known exactly. If the same characteristic is repeatedly measured with an "accurate" device under constant conditions, the same result will not always be obtained. As a matter of fact, the same result will seldom be repeated. It is assumed that these observed deviations result from chance errors in the measuring process. To cope with this inherent deficiency, repeated measurements are made from which an interval is calculated, which we believe includes the true value represented by the data.

Confidence Interval

Intervals of the kind discussed above are called confidence intervals. Included in the method of calculating these intervals is a means of controlling the proportion of the time that the true value is expected to fall within the interval. This proportion expresses our "confidence" of being right in our prediction that the true value will fall in the interval calculated.

POPULATION VERSUS SAMPLE

The family of values generated by repeated measurements of the same characteristic is called a *population*. A population is generally assumed to be infinite. Any subset of a population is called a *sample* of that population. A sample is always finite.

PREDICTION

The observations of measurements made in any experiment are finite samples of a much larger (infinite) body of data that could exist if thousands (infinite) of observations had been made of the same characteristic under the same constant conditions. It is assumed that, unless an infinite number of observations is made, the true value of the characteristic measure will never be known exactly. This reasoning requires focusing of attention not on the *observed values* but on *what these values represent*—the larger family of all possible values of the characteristic being measured. The objective is to infer from the sample something about the population. Experience has taught that prediction (an inference) cannot be made with certainty. There is always a chance of being wrong. Errors of this type are called *prediction errors.*

FREQUENCY DISTRIBUTION

If all measurements referred to above are divided into small groups or cells having a range equal to about one-tenth the total range (from maximum to minimum) of all values, there will be about ten cells.

Then if a count is taken of the number of values falling within the range of a particular cell, the ratio of this number to the total number of measurements available is the relative frequency of occurrence of measurements (events) in that cell. If the total number of measurements available is very large (1000 or more) and *all* values falling within the cell are counted, a very good estimate of the true frequency of occurrence of values in that cell for that particular population will result.

Normal Distribution

As the total number of values used is increased and the cell width (range) decreased the stepwise form of the histogram fades into a smooth curve that is called a frequency distribution. This is actually how a frequency distribution is formed in practice. It means what the name implies; it is a distribution of (relative) frequencies. Experience has shown that the families of values generated by repeated measurements of the same characteristic under controlled conditions have definite forms. The most common of these forms, and the most useful, is called the *normal* frequency distribution or bell curve. The family of values forming this distribution is called the normal population.

As the cell's width in the histogram decreases and approaches zero, the height of the bar represents the relative frequency for a single value on the abscissa. Thus there is a relative frequency for any value in the population of the measurements. The sum of all the frequencies equals the frequency of all values in the population. This is assigned the numerical value of one. The equation for this function is known, but it is no direct importance to this discussion. It can be found in any standard text on statistics.

Probability

From a practical point of view, relative frequencies (proportions) are estimates of probabilities. By definition, if an event is certain to occur, the probability of occurrence is said to equal unity. If it is certain that an event will *not* occur, it is said that the probability of occurrence is equal to zero.

In the above example, if the cell width was equal to the range of the population (from the maximum value in the population), it would be certain that the next value taken would fall within this "cell." As a result of taking repeated measurements, *all* the values would fall within this cell. The number of values falling within this cell divided by the total number of values will equal unity. That is, the probability of a value's falling within the cell (the event) is equal to one.

If, on the other hand, a new cell is taken that has a maximum limit less than the minimum of the above population, it is certain that the next value taken from the above population will *not* fall within the new cell. If repeated measurements are taken from the above population, *none* of the values will fall within the new cell. The number of values falling within the cell divided by the total number of values will equal zero. That is, the probability of a value falling within the cell (the event) is equal to zero.

The area under the normal frequency distribution is used to measure probabilities. As shown above, the magnitude of the ordinate associated with any value on the abscissa is a measure of the relative frequency of occurrence (or probability of occurrence) of that value. The summation of all the ordinates below any particular value on the abscissa is, of course, equal to the area under the curve below that ordinate. This area is then a measure of the probability of occurrence of all values in the population below the given value.

Parameters

Just two parameters or characteristics of the normal frequency distribution are required to define this curve completely. The first parameter is the central value around which most of the values belonging to a particular population will naturally cluster. This parameter, called the true or population *mean*, is measured by the arithmetic average of *all* the values in the population. The other required parameter is the dispersion of values around the central value. This parameter is measured by the root mean square of the deviations from the true mean and is called the true or population *standard deviation.*

Properties of Normal Curve*

Graphically, the mean is the ordinate that passes through the center of gravity of the area under the curve, since this curve is symmetrical. The mean is equal to the mode (the most frequently occurring value) and the median (the middle value).

Also, the standard deviation is equal to the horizontal distance between the ordinate of the mean and the inflection point on the curve on either side of the mean ordinate.

The true mean plus or minus one standard deviation includes 68.27% of the total area under the curve. The mean plus or minus two standard deviations will include 95.45% of the total area under the curve. These values are used to make probability statements. They mean either or both of the following:

1. In generating a normal family of values, 68% of the total number of values will lie within plus or minus one standard deviation of the mean. This is especially true if the total number is very large—that is, 1000 or more.
2. Randomly chosen values from the normal population have a 68% chance or a probability of 0.68 of falling within plus or minus one standard deviation of the mean.

This distribution is unique in nature. It is the curve of regression for the distribution of all small sample averages.

Randomization

The meaning of the word *random* as used in statistics, can be better described than defined. The phrase "randomly chosen values" describes a selection procedure of a very special kind. This procedure is free of biases of all sorts. It is the only procedure that will permit the free play of chance variations, which are the theoretical basis for all modern statistical techniques.

Random selection or random sampling can be accomplished by physically mixing the items before sampling, or by numbering all of the items and then using a table of random numbers to determine which items to select and in what order to select them. Random selection is the process used in lotteries; all numbered tickets are deposited in a revolving drum and a single drawing is made by a blindfolded person. It is assumed that such a procedure is completely unbiased, that chance

---

* Refer to *Recommended Readings.*

alone is at play, and that each ticket in the drum has an equal chance of being selected. The process of random selection, then, not only permits the laws of chance to determine *which* item is to be chosen, but also the *order* in which successive items are chosen. This procedure relieves the PT&E evaluator completely of any responsibility concerning "which item" and "which order" in the selection process. In an experiment, the experimenter wants to be unbiased.

Sampling

In a lottery the relative frequency of occurrence of any particular number is equal to the relative frequency of occurrence of any other number in the drum (population). Each number has an equal chance of being selected if the selection procedure is truly random and unbiased. However, the relative frequency of occurrence of the values in a normal population is not equal. Theoretically, all are different. Reflection will show, however, that random selection will be "fair" and unbiased here also. If 900 white beads and 100 red beads are mixed well in a bowl (that is, randomized), and a handful of beads is selected by a blindfolded person, the ratio of red to white beads in that handful (or any other handful) will be approximately one-to-nine. The average ratio of a large number of trials (handfuls) will be one-to-nine—the relative frequencies of the two colored beads in the bowl, the population. The same relation between sample and population holds true in selecting (sampling) values from a normal distribution if sampling is done in random fashion. That is, every value (or item) in the population has a chance (probability) of being selected equal to the frequency with which it actually exists in the population. Only samples that can reflect these actual relative frequencies in the population can be considered as representing the population from which they are taken if valid inferences are to be made. Of course, successive samples drawn from the same population will not be identical but, if randomly selected, the difference between them will result from chance errors only. Under these circumstances, modern statistical techniques will identify them as having come from the same population which, in fact, they did.

Estimates

In practice, to make a measurement (or observation) is to estimate the true population *mean*. The more observations made and averaged together, the better the estimate. This estimate is called a "point estimate" to distinguish it from an "interval estimate." However, it is assumed that the true mean is never known exactly unless an infinite number of observations are made.

If the root mean square of the deviations of the individual observations is calculated from the average of all observations the true population *standard deviation* can be estimated. As with the mean, however, it is assumed that this true parameter is never known exactly unless an infinite number of observations are made.

Predictions

The two predictions made most often in statistics are the following ones:

1. The magnitude of the true parameters. These predictions are based on interval estimates, which are called confidence intervals.
2. Whether two or more values belong to the same population. These predictions are called tests of significance.

The prediction problem in statistics is to estimate the population mean and standard deviation and then predict what these two population parameters might be or, given two or more estimates, to predict whether they came from the same population. If there are thousands of observations in each of the samples, the sample means and standard deviations are equal to the population parameters, and prediction becomes unnecessary. In practice, however, such large samples are generally not available because they are too costly to obtain. The problem, then, is to predict from small samples what the parameters might be or whether the samples came from the same population.

Intuitively, it is known that predictions cannot be made with certainty—there is *always* a possibility of being wrong. As a result, to be right as often as possible, reliance is placed on planning. Through statistics, this possibility is maximized, and the chances of being right are actually controlled.

On a mathematical basis the data are assumed to have a normal frequency distribution. The normal distribution is then used to calculate the probability of being right in making predictions. This is called the *confidence level* of predictions. The assumption that the distribution is normal for variable (measured) data is a reasonable one. Experience has shown that the numerical values of measurable characteristics of products manufactured under controlled conditions are normally distributed. In addition, the central limit theorem states that the distribution of averages of variable data is normal. Thus, in comparing averages or calculating confidence intervals of measured data, the assumption of normality is quite valid. Notice that this is true of attribute data only where they are converted to variable data by some process such as the arcsine transformation.

Test Design Considerations

Certain requirements must be fulfilled in order to prepare a good test or experiment. Frequently, in the course of day-to-day activities, these factors are overlooked. Specifically, the factors requiring early evaluation include scope, test hypothesis, performance criteria, variable selection, factor levels, measurement type, treatment selection, and specimen selection. They are discussed below.

Scope. Consider the entire scope of the problem. Without regard to cost, time, or effort, consider what must be known eventually. If this turns out to be a very large experiment at a prohibitive cost, divide the whole problem into rational parts. This makes possible a systematically planned approach and a tie-in of your test plan to cost and the amount of information required.

Test Hypothesis. Develop the right hypothesis by asking the right questions that the experimental results are expected to answer. Conclusive proof that component A has a much higher reliability than component B solves nothing if component A cannot be consistently produced.

Performance Criteria. Choose carefully the criteria on which conclusions will be based. Insistence that a component must have a particularly high performance requirement with respect to one parameter is of little value when it has only marginal performance with respect to some other essential parameter.

Selection. Make a comprehensive list of all the variables (factors or environmental treatments) whose effects on the component's functioning characteristics are of interest or must be determined. This should include factors of direct interest, factors that may help to show how the main factors work, and factors required to determine the effect of experimental technique. In addition to choice of the variables to be used, their order of use must be established. The order chosen should be the one most likely to be experienced in use or the one considered most severe, and must be held constant throughout the experiment.

Factor Levels. The number of levels used in the designs described here is limited to two. These designs are the simplest and the most versatile for conducting multifactor experiments.

These two levels are usually the extremes, such as the presence and absence of an environmental treatment or extremely low and high temperatures. The choice of levels must be arrived at through good judgment, common sense, and detailed knowledge of the purpose and probable outcome of the investigation. Factorial experiments are most efficient in their ability to detect differences among environmental effects when the levels of severity used are such that approximately 50% of the test specimens fail.

Type. The type of measurement to be used should be considered for the sake of efficiency. Variable types of data can vary from minus infinity to plus infinity and can furnish the maximum information per observation. Attribute data are "success" "failure" type of data and furnish the least information per observation. From this, it is clear that variable type of data should be used wherever possible in factorial experiments.

Treatment Selection. Treatments are chosen to give as direct an indication as possible of the functioning characteristics of the components, and to include as many as possible of the environmental conditions expected in use. This is an engineering decision that must be based on good judgment consistent with the purpose of the experiment.

Specimen Selection. A test specimen is the smallest subdivision of the experimental material that can receive different treatments. Sufficient homogeneous or uniform material should be available to conduct a complete set of treatment combinations (required by the experimental design) during a single period of time (such as a day) by a single instrument condition (such as calibration) and by a single operator or group of operators. Material produced during a particular period of time by a single process and by a single manufacturer normally can be considered homogeneous.

The experimental units used should not differ in any important respect from the best known unit design to which the conclusions are to apply. If design changes are made on the basis of experimental results, the units used to obtain the results are, of course, not representative of the modified design.

Experimental units should respond independently of one another. Obtaining a failure on one should not affect any of the others. Using a separate item for each treatment combination will usually assure independence.

Progressing now into the actual design of the test program, certain other factors become important from a statistical and PT&E point of view. Among these are the design selection, sample size, randomization, and replication.

The factorial design and its modifications, described later in this chapter, meet the requirements of environmental testing experiments better than any other known design. The advantages of the recommended factorial designs for environmental testing are as follows:

1. Simple to use and analyze.
2. No control groups are required.
3. The two levels of each treatment can be the presence and absence of

the treatment, if desired. Alternatively, any two levels of the treatments can be used.

4. Each treatment effect can be determined independently of all the others. Unambiguous conclusions can be drawn about the effects of each treatment.
5. Complex experiments, involving a large number of treatments, can be easily handled.
6. These are the only experimental designs with which the relationships among treatments can be measured. These designs can determine whether the effect of one treatment depends on any of the others. These relationships are called interactions.
7. The probability of being right or wrong can be controlled.
8. When the number of treatments used becomes large (three or more), only a fraction ($1/2$, $1/4$, $1/8$, etc.) of the total number of combinations of treatments and levels needs to be used. These designs are called fractional factorials and optimum multifactorials.
9. A type of statistical analysis can be used that distinguishes between variations that result from chance and variations that result from assignable causes.
10. More information can be obtained from a given number of test specimens than from any other known procedure.
11. The effective sample size is increased by making it possible to use each observation (or measurement) for more than one purpose.

In fact, each treatment effect is determined as though the entire experiment is conducted to determine that particular treatment effect alone. As a result, the precision with which each treatment effect is determined can be based on the total number of test specimens used in the experiment.

TEST SPECIMEN BALANCE

In any PT&E experimental situation a reasonable balance must be established between using too few test specimens, thus obtaining poor precision, and wasting time and material in attaining unnecessarily high precision by using too many test specimens. When a preassigned number of test specimens is available, the question arises: Is the experiment worth doing at all? If the number of test specimens available is flexible and adequate, the number required for a given precision or reliability can be calculated in advance. The minimum number of test specimens required in the optimized designs is only one more than the total number of treatments used. The more versatile factorial designs require at least 16 items for 5 to 8 treatments, and at least 32 items for 9 to 13 treatments. With twice these numbers of items, the latter designs can also measure interactions.

ORTHOGONALITY AND CONFOUNDING

The property of these designs, known as orthogonality, must be preserved in order to simplify the analysis and interpretation of the results. This can be done by keeping the number of observations per treatment combination equal and constant throughout the entire design. Orthogonality assures that all of the environmental effects and their interrelationships can be independently estimated without entanglement.

Confounding is the converse of orthogonality. It means equating two or more factors or treatments so that their separate effects cannot be determined. For example, little can be concluded about the separate effects of the environmental treatments if all the treatments are applied to each item. If a failure is obtained after an item has received two or more treatments, the cause of the failure is ambiguous; it could result from any of the following:

1. The last treatment.
2. The last two treatments.
3. All the treatments.
4. Any of the other possible combinations.

The exact cause cannot be determined because the treatments are confounded. This type of confounding should be avoided. Interaction is said to be present when certain particular treatment combinations produce unusual results. This is the nonadditive or unpredictable portion of the experiment. When appreciable interaction effects are present, care must be taken in quoting main (average) effects. Any statement about the average effect of treatment must specify the level of the interaction.

However, the determination of interaction effects may be the most important information obtained from an experiment. It can explain what otherwise appear to be contradictions. This is extra information furnished by factorials that cannot be obtained from other designs. Plans should be made to use factorials that can measure interaction effects if there is a possibility that they exist. Higher-order interactions can be used as estimates of the error term when multiple replication is not used.

BLOCKING

In general, blocking means dividing the entire design into orthogonal subgroups. This reduces the number of observations that need to be taken in one continuous period of time and reduces the amount of homogeneous material required in one batch. Difference among blocks due to uncontrolled changes with time and due to changes in material can be mathematically subtracted out of the system. That is, the object of blocking is to make it possible to conduct the experiment in reasonably small portions. Plans should be made to block any large experiment or any experiment expected to extend over a long period of time. Taking observations in complete replication sets is one form of blocking.

RANDOMIZATION

Randomization can be accomplished by means of a table of random numbers or by drawing well-shuffled numbered cards from a hat. The important characteristic of randomization is that it can be an objective impersonal procedure. Proper randomization is determined by examining the procedure producing it, not by examining the results. To randomize does not mean to arrange in an order that looks haphazard. The object of randomization is to permit the laws of chance (probability) to operate freely. Proper randomization is the most important requirement for a good experiment because it prevents biased results of all kinds due to such factors as human prejudice, weather cycles, trends in time and heterogeneity of experimental material.

However, randomization can be abused. Randomization should not be used in PT&E to conceal large variations. This drastically reduces the sensitivity of the experiment to detect small differences. All variables known to have, or suspected of having, significant effects on the outcome of an experiment must be either controlled or designed into the experiment. Randomization should be used only to remove the effects of small variations after every other source of variation has either been included in the design or controlled. Only good engineering judgement and a knowledge of the system can determine how, when, and where to use randomization.

REPLICATION

Replication means repetition. One complete replication consists of a single observation for each of the treatment combinations in the design. If the observations are performed in sets so that a complete replication is done in a continuous period of time (such as a day), with a single measuring system (or instrument) by a single operator, the difference among replications can be used to determine whether the external experimental conditions have remained under control. Multiple replications are also used for the following purposes:

1. To increase the precision with which treatment effects are determined.
2. To furnish an independent measure of the error term.
3. As a basis for calculating the failure rate observed for each treatment combination in preparation for transforming attribute data to a continuous scale in analysis of variance procedures.
4. To remove systematic error.
5. To relieve the experimenter of the responsibility of choosing which item of test or which test to conduct. Each item or test is equally likely to be chosen. In this sense the experiment is "fair" and unbiased.
6. To assure the validity of statistical techniques, such as the analysis of variance and associated tests of significance which depend for their validity upon the laws of probability.

The final step in the process is the analysis of results. This step is greatly facilitated if the considerations previously mentioned have been adequately considered earlier.

STATISTICAL SIGNIFICANCE

The word "significance" has a special technical meaning in statistics. Its meaning must be understood in order to analyze statistically and to interpret experimental results. One of the most important contributions of statistics is that it has established a means of distinguishing between chance variations and assignable causes. When the observed differences result from chance variations, these differences are said to be nonsignificant. This means that the observed results originated from the same source (population). When the observed differences have assignable causes they are said to be significantly different. This means that the observed results have originated from different sources (populations). In a well-planned experiment these sources can be identified. For a nonsignificant difference, changing the treatment from its lower level to its higher level will not cause a detectable difference. On the other hand for a significant difference, changing the treatment from its lower level to its higher level will cause a detectable difference.

Treatment Effects

In a properly designed test for PT&E, each treatment effect should have a unique interpretation. If two or more interpretations are possible, additional work is required to clarify the ambiguities. One of the most important requirements of a good test design is that the conclusions be unambiguous. Fortunately, factorial designs are very helpful in avoiding ambiguity. To conclude that an effect is not significant is not the same as saying that the effect does not exist. We can only say that there is insufficient data to detect the effect. However, if the conclusions are that the effects are significant, we can be assured that the effects are real to the extent of the confidence level associated with the test of significance. Further advantages of factorial designs are as follows:

1. The range of validity of the conclusions concerning the average (main) effects is extended by the inclusion of more than one variable in the experiment.
2. Physical interpretation of interactions explain and clarify underlying mechanisms and relationships.

When only one observation is taken for each treatment combination, analysis of the results from the factorial designs described here is simplified by using tables of minimum contrasts.[1] These tables are based on the binomial distribution. The test of significance that uses the values in these tables is known as "Fisher's Exact Method for 1 × 2 Contingency Tables."[2] This test is valid even for small-sample sizes, and will determine not only the main effects but also the two-factor interaction effects when the proper designs are used. When multiple (but an equal number of) observations are taken for each treatment combination, the Fisher method can still be used. However, an alternate method, which is slightly more efficient but which requires more calculating, can also be used. This method transforms the qualitative data to a continuous scale through the use of arc sine of the proportion or percentage of failures found for each treatment combination. The transformed data can be analyzed by the usual analysis of variance techniques. The tests of significance and their interpretations are both made by using the transformed data.

For quantitative data (such as g-values, voltages, or time), the usual analysis of variance can be conducted on the observed data, provided that the variances are homogeneous throughout the design. Since this procedure is somewhat involved, lengthy, and adequately covered in literature,[3,4] variance techniques are not analyzed here.

FACTORIAL TEST DESIGNS

Factorial designs for PT&E have been selected for discussion because of their preferred value over other test designs. Notice, however that the binomial distribution is assumed to be applicable to use these methods. The two-to-the-nth factorial designs, or their optimized modifications, are the most efficient experimental methods known for selecting the treatments that cause the highest failure rates. This approach will reduce the magnitude and complexity of the experiments required to determine product performance. More important, all component data obtained in this manner will have a common basis of interpretation because the performance of each component is defined in terms of the environment that has been experimentally found to cause the highest failure rate.

Experiments based on these designs may be conducted without changing the treatment procedure, except to arrange for the test specimens to receive the number and kind of treatments required by the particular design used. However, the best differentiation among treatments is obtained when the level of severity used will cause 50% of the test specimens to fail. This may cause some adjustment of the levels of the treatment used.

For the purpose of this application, the two levels of each treatment can be the *presence and absence* of treatment. Alternatively, any two levels of the treatment can be used.

The number of test specimens required in the optimized designs is one more than the total number of treatments used.[5] The more versatile fractional factorial designs[6] require at least 16 items for experiments containing from 5 to 8 treatments, and at least 32 items for 9 to 13 treatments. With twice these numbers of items, the latter type of designs can also measure interactions (that is, how the effect of any one environment depends upon the others). Interactions among treatments cannot be measured except by factorially designed experiments.

Factorial designs for PT&E permit a type of statistical analysis that distinguishes between variations that result from chance and variations that have assignable causes, thereby producing more information from a given number of items than any other known procedure. These designs actually increase the effective sample size by making it possible to use each observation (or measurement) for more than one purpose.

In fact, each treatment effect is determined as though the entire experiment is conducted to determine that particular treatment effect alone. As a result, *each* treatment effect is determined with a precision equal to the total number of items used in the experiment. When multiple replications cannot be used and only attribute (go, no-go) data are available, these designs can still be used to take advantage of their efficiency. However, in cases of this kind, the usual analysis of variance cannot be made. Instead, the usual summations are made to obtain and compare two binomial proportions (by the Fisher exact method) to determine the effect of each treatment.

FULL FACTORIAL DESIGNS[3]

These PT&E designs require more specimens per treatment than do the fractional factorial designs, but they are the only class of designs that can measure *all* of the interaction effects. A full factorial can be formed by writing down all of the combinations of n treatments, each at two levels in a multientry table. For example, a full two-cubed factorial can be written as shown in Table 1.

The lower-case letters and the symbol (1) in the body of the table identify each of the eight ($2^3 = 8$) treatment combinations that constitute this design. These combinations are derived from their position in table. For example, the symbol (1) is located by $A_1B_1C_1$, which means that all three treatments are at their higher levels and that treatment B is at its lower level.

In this code the lower case of the treatment letter appears in the combination only when the treatment is at its higher level. This results in the information of all possible combinations of n things (treatments) taken 0, 1, 2, ... and n at a time. At least one test specimen or observation is required for each of the treatment combinations. Two or more observations at each treatment combination are required for an independent estimate of experimental error.

An equal number of observations at each treatment combination is required to keep the design orthogonal.

FRACTIONAL FACTORIALS[6]

As the number of treatment variables increases, the number of treatment combinations (and, therefore, the number of test specimens required for a complete replication) increases very rapidly. At the same time the number of higher-order interactions that can be measured also increases very rapidly. This results in two undesirable situations:

1. The number test specimens required becomes excessive.
2. The information in the higher-order interactions (three-factor interactions and above) is of little practical use. Fractional factorial designs were developed to avoid these situations and thereby improve the efficiency of designs for multifactor experiments.

When less than all of the possible combinations in a factorial design are used, the design is said to be a fractional factorial. For the two-to-the-nth series, there can be half, quarter, eighth, sixteenth, etc. portions of the full factorial used. These portions are called fractional replicates, where a full factorial is one replicate.

Table 1 2^3 Factorial Design

	A_1		A_2	
	B_1	B_2	B_1	B_2
C_1	(1)	b	a	ab
C_2	c	bc	ac	abc

Fractional factorials cannot be used without losing or giving up some information that is available in the full factorial. However, it is planned in designing a fractional factorial to lose only the least important part of the information. Experience has shown that the higher-order interactions in a full factorial are the least important. This fact is utilized by equating new treatments to the higher-order interactions. The equation of one such interaction to a new variable in a full two-to-the-sixth factorial, for example, creates a half replicate of a two-to-the-seventh factorial. A detailed procedure for designing fractional factorials can be found in the reference cited in a footnote 3. At least one observation for each treatment combination is required to keep these designs orthogonal.

An example of one type of analysis that can be used with factorial designs follows this paragraph. This is the simplest possible analysis. The type of analysis that can be made depends on the class of design used, the kind (attribute or variable) and amount (number of replications) of data, and the way (at random or in blocks) that data were collected. Some types of analysis, such as the analysis of variance, are quite complicated. As a result, the subject of the analysis of variance (see footnote 3) is not included here.

EXAMPLE OF FACTORIAL EXPERIMENT

This example demonstrates how factorially designed environmental experiments can be used in PT&E. A simple three-treatment-experiment example is given below. The treatments use in this example are identified and defined as follows:

Identification	Treatment
A	Transportation vibration
B	Powered flight vibration
C	High temperature

For purposes of the factorial design, each treatment is considered to have two levels:

1. Lower level is the absence of the treatment (designated by subscript 1).
2. Higher level is the presence of the treatment (designated by subscript 2).
3. The total number of possible combinations of three treatments, each at two levels, is two cubed or eight. These eight combinations can be written in the following pattern:

$$
\begin{array}{ccccc}
 & & A_1 & & A_2 \\
 & B_1 & B_2 & B_1 & B_2 \\
C_1 & (1) & b & a & (a+b) \\
C_2 & c & (b+c) & (a+c) & (a+b+c)
\end{array}
$$

A minimum of eight items would be required for this plan, each receiving different treatment combinations as follows:

Item	Treatment combinations
1	None (1)
2	B only
3	A only
4	A + B
5	C only
6	B + C
7	A + C
8	A + B + C

A knowledge of the item being tested has led to the decision that transportation vibration, powered flight vibration, and high temperature, *in that order*, are the three environmental conditions most likely to affect the important functioning characteristic of this item. The treatment procedure and assumed results for this experiment are as follows. A plus mark in the item column means that the item received the corresponding treatment, while a "blank" means that the item did not receive the treatment.

Interpretation (When the Above Order is Used)

1. The replication effect is not significant. This means that the conditions of the experiment did not change significantly from beginning to end. Therefore, the results can be accepted as valid from this point.

2. None of the effects is significant except the $A \times C$ interaction. This means that the combination of transportation vibration and high-temperature treatments has caused a larger difference in the number than would be expected due to chance variations alone.
3. None of the treatments taken alone is significant, although the powered flight vibration and transportation effects approach significance. These results suggest the need for additional flight vibration and transportation vibration tests if these treatments are considered important from an engineering point of view.

These results suggest the need for additional flight vibration and transportation vibration and high temperature is the most severe condition.

SAMPLE SIZE

Because of economic considerations, the question of how many specimens to test (or how large a sample size to use) is always given a prominent part in planning any testing program. It is the question most often asked by engineers concerning testing programs. An answer from only the economic point of view is not enough because the cheapest testing program is none at all: of course, if no testing is done, there is no verification that the newly developed item is usable or not.

Before the question of sample size can be answered, the following related points must be considered:

1. The notion that reliability is related to the number of specimens tested must be discarded. Only the *precision* with which the reliability is determined is related to the sample size.
2. There is no one single sample size that is applicable to all testing programs. Each program must be considered individually.
3. A valid sample size cannot be stated without first knowing the *purpose* of the testing program. It is very easy to get the right answer to the wrong question.

The purpose of planning the sample size prior to data collection is to obtain essential information (the minimum information required so that additional data will not change the conclusions) with minimum cost, effort, and material. To accomplish this, the following design of experiment techniques must be considered, since the question of sample size cannot be answered outside this context. (1) The purpose of any testing program is to verify the hypothesis that objectives have been achieved or maintained. (2) To do both, the sampling and testing procedures must achieve maximum precision with minimum sample size.

In all practical testing programs (especially those in which the testing is destructive), something less than all the existing items should be tested and, from this, an inference should be made about the remaining (usually larger) portion of items. To have these inferences valid, the sample must "represent" the remaining portion of the lot or population. If the lot is homogeneous, a representative sample can be obtained by random selection. That is, each individual item in the lot must have an equal chance of being selected. If the lot is not homogeneous but is stratified in some manner according to geographical location, or manufacturing process, then the sampling plan must be designed to cope with this characteristic. If the strata are only few in number, then an equal number of randomly selected specimens should be selected from each stratum. The number selected should be apportioned according to the size or importance of each stratum. If the number of strata is large, then specimens should be taken from only a part of the strata. If all strata are equivalent, then a sample of the total number of strata should be randomly selected before the specimens within them are randomly selected. If all the strata are not equivalent, then the most important or largest strata should be used. In any case, the strata or specimens must be selected randomly either by physical mixing and selection, or by numbering and determining which numbers to select with a table of random numbers.

It is important that the sample be stratified correctly to parallel that of the lot. Only in this way can the heterogeneity of the sample be kept to a minimum. Any increase in the heterogeneity of the material, due to sampling, results in an inflation of the overall variation as measured by the standard deviation of the testing method. As shown below, the magnitude of the standard deviation is of prime importance in calculating sample size.

TEST MEASUREMENTS

The testing method is, in reality, a measuring system or device that must be precise. It "measures" the characteristic of the item being used as a basis for evaluation and decision. By precise is meant that characteristic of the method that produces estimates (numerical results) from repeated trials that are close together when, in fact, there has been little or no variation in the system. Methods that produce estimates close together and do not reflect variations that actually occur in the system are called insensitive rather than precise methods. Such methods are to be avoided. The sample size varies directly as the square of the standard deviation. The only way that precise methods can be made available is via a continuous program of development.

Because all testing is done on a sample basis, decisions (inferences) must be made about the lot based on the information gained from the sample. This, actually, is a form of prediction, and predictions cannot be made with certainty. However, only two kinds of error can be committed in drawing inferences about the lot: (1) Type I error is rejecting good material, (2) Type II error is accepting poor material.

It is desirable to keep both of these errors small. Their magnitude can be controlled by the number of specimens (sample size) tested in any given situation, as shown below. In practice, the magnitude of these errors chosen is based on the consequences of being wrong. [For example, the consequences of rejecting good material (Type I error) can cost only dollars, but the consequences of accepting poor material (Type II error) can cost lives and lose customers. Therefore, the sample size required to maintain both errors at the selected levels is calculated.]

In any testing program a decision must be made on at least one of the following requirements before anything can be said about sample size.

1. The maximum confidence interval that can be tolerated for the particular purpose intended.
2. The minimum difference (between two values) necessary to be detected for the purpose intended.

These requirements can be established only through knowledge of the objectives and purposes of the system under consideration. Fortunately, this kind of information is usually well known. Sample size varies inversely as the square of the difference to be detected. The sample size required to detect a difference of $(d/2)$ is four times that required to detect a difference of (d).

If the effect of more than one variable (such as the effect of more than one environment) must be determined, test design is extremely important in keeping sample size to a minimum. If test treatment is correctly chosen, the efficiency of the experiment can be greatly enhanced. In fact, the efficiency is improved by a factor equal to the number of variables included in the design. For example, to obtain a given precision, a factorial design for three variables requires only one-third the number of test specimens required by the classical one-at-a-time procedure. Factorial design for seven variables requires only one-seventieth the number of test specimens required by the classical one-at-a-time procedure. At least one test specimen is required for each combination used. When this number becomes large, only a part of the total number of combinations need be used on designs called fractional factorial designs. Fractional designs have the same high efficiency as the full factorial designs, and should be used to screen all the variables of interest to find the most important ones—such as the most severe environment.

HYPOTHESES TESTS

Tests of hypotheses are used to compare two or more performance values. The purpose of tests of this kind is to determine whether they result from assignable causes. This determination is important in decision making. To decide that the performance value obtained during the second testing period is lower than that obtained during the first testing period is very disconcerting, if the value obtained in the third testing period is higher than the first value obtained. This is especially disturbing if the decision has led to more testing or replacement of parts, but this is exactly what can happen if the observed differences result from chance variations. Only through use of statistical tests of significance can this difficulty be avoided.

In hypotheses, testing both the alpha (Type I) error and the beta (Type II) error is especially important because of the consequences of being wrong. The only way that these errors can be controlled at predetermined values is to calculate the sample size required to do so in advance of data collection. Experience has shown that when these two kinds of errors are kept at 5% or below, the risk of making a wrong decision is sufficiently low for most purposes. To reduce these errors below 1% requires very large sample sizes.

For lots made up of discrete items from which only attribute (success or failure) data can be obtained, the following should be considered:

1. If the lot is finite in size and less than ten times the size of the sample selected, then this fact must be taken into account. The action taken in this regard depends upon the purpose of the testing and the scope of the conclusions drawn as described below.
2. If the testing done destroys the specimens selected or if the specimens are not returned to the lot for any other reason, this fact must also be taken into account. Again, the action taken depends on the purpose of the testing and the scope of the conclusions drawn as described below.
3. A decision should be made prior to data collection concerning the purpose of the testing. If the purpose is to draw a conclusion about only those items in a small (finite) lot and if both of the above two conditions pertain, then the sample size can be reduced slightly through use of the hypergeometric distribution. This distribution

finds its most frequent use in acceptance testing where the purpose is to predict the expected fraction defective of a particular small lot of items. If the small lots of material in the pipeline represent larger lots of indefinite size, the characteristics of this material are studied and recorded for their value in future applications. That is, the purpose of testing is to draw inferences about the larger volume of material represented by the small lot on hand. In this latter case, only the binomial assumption concerning lots of infinite size is applicable.

CALCULATION

Sample Size Required for a Given Confidence Interval
Variable Data

$$n = \frac{(ts)^2}{d^2}$$

where

n = sample size
t = standard deviation associated with the alpha error used to control the confidence level
s = sample standard deviation
d = magnitude of the confidence interval in the same units as the standard deviation

Attribute Data

1. *Binomial distribution.* There is no easy, practical way to calculate accurately the sample size required for attribute data. The accurate methods are difficult to calculate and the simple, easy methods are not accurate. The most practical method is to refer to one of the existing tables for binomial confidence intervals to find the sample size required for a given interval.

2. *Hypergeometric distribution.* As with the binomial distribution, there is no easy, direct way to calculate the sample size for the hypergeometric distribution. The most practical way to arrive at a sample size in this case is to refer to one of the existing tables for the hypergeometric confidence intervals. From these tables the sample size for a given interval and confidence level can be read directly.

Alternatively, the sample size required in a hypergeometric distribution can be estimated by multiplying the sample size for the binomial distribution by

$$N \div (N + n)$$

where

N = lot size
n = sample size required in the binomial distribution

Sample Size Required to Detect a Given Difference between Two Sample Values in Testing Values in Testing a Hypothesis:
Variable Data

$$n = 2\frac{(t_1 + t_2)s^2}{d}$$

where

n = sample size
t_1 = standard deviation associated with the alpha error
t_2 = standard deviation associated with the beta error
s = sample standard deviation
d = difference that must be detected

Attribute Data. As mentioned above, there is no easy practical way to calculate the sample size for attribute data. The sample size for hypothesis testing, using attribute data, can best be determined from existing tables for minimum contrasts.[2]

In this table, N is the sample size. The values in the body of the table that appear to be proportions are written in a shorthand method which means the following:

For a sample size of five, the value in the first column of this row (0/4) means that if no failures are obtained in the first sample of five, at least four failures must occur in the second sample of five before the observed difference can be declared significant at the 95% level of confidence. This, in turn, means that a sample size of five can only detect differences of 80% or greater. Larger sample sizes can detect smaller proportional differences. The use of these tables can, of course, be reversed to find the sample size required to detect a given difference.

SUMMARY

This chapter provides, in abbreviated form, some of the more important statistical considerations relevant to product testing and evaluation. Basic statistical ideas were summarized and related to specific applications so that the reader should be able to relate the importance of proper test design to the economic and technical value of a given test program. Extensive use has been made of existing statistical concepts without derivation. The recommended readings suggested below will provide the interested reader with necessary details.

RECOMMENDED READINGS

Brownlee, K. A. *Industrial Experimentation.* 4th ed. New York: Chemical Publishing Co., 1953.

Grant, Eugent L. *Statistical Quality Control.* 2nd ed. New York: McGraw-Hill, 1952.

Ostle, Bernard. *Statistics in Research.* 2nd ed. Iowa: The Iowa State University Press, 1963.

NOTES

[1]D. Mainland, L. Herrea, and M. I. Sutcliffe, *Tables for Use with Binomial Samples,* (New York: Department of Medical Statistics, New York University College of Medicine, 1956).

[2]R. A. Fisher, *Statistical Methods of Research Workers,* (New York: Hafner Publishing Co., 1950).

[3]O. L. Davies, (editor), *Design and Analysis of Industrial Experiments,* (New York: Hafner Publishing Co., 1954).

[4]Henry Scheffe, The Analysis of Variance, (New York: John Wiley & Sons, 1959).

[5]R. L. Plackett, and J. P. Burma. The Design of Optimum Multifactorial Experiments, *Biometrika*, Vol. 33, (1946) pages 305−25.

[6]*Fractional Factorial Experiment Designs for Factors at Two Levels,* NBS Applied Math. Series No. 48.

Chapter 11

Product Differentiation Procedures

A manufacturer may effectively differentiate his product in the market-place on the basis of reliability via the evaluation technique of screening tests. In this way he is able to claim the premium price commended by the higher product quality associated with his product. He has in effect segmented his market based on product performance characteristics which are tied to quality.

The purpose of reliability screening is to select, from a set of devices, the ones that have superior reliability or, alternatively, to reject the ones that have inferior reliability. A particular screening problem of wide interest is the screening out of component parts that are potential early failures in some specified application. These screening procedures involve the classification of each device in the set on the basis of initial or early life parameter measurements. Depending on device reliability requirements and the "power" of the screening procedure, devices may be screened into two or more classifications. For example, in screening sets of component parts, one may wish (1) to eliminate the potential early failures, (2) to select the parts exhibiting high reliability for appli-cation in life support systems, and (3) to select the parts of intermediate reliability for commercial applications with demanding life requirements.

Reliability screening differs from quality control in several respects. First, it is not the purpose of reliability screening to detect devices that are defective at the time of measurement. In fact, it is assumed that all devices are initially good. Second, unlike acceptance sampling, parts qualification, and the like, reliability screening is a 100% inspection procedure. Third, classification by a screening procedure is accomplished with respect to lifetime requirements and operating conditions involved in the intended application of the device. Thus the concept involved in reliability screening, as opposed to quality control, is not to reject a device strictly because one or more initial parameter measurements lie in some unacceptable region but, instead, to consider such measurements as precursors of early failure (unreliability).

Ideally, a screening procedure would never misclassify a device. That is, a high-reliability device would never be classified as a potential early failure and, conversely, an unreliable device would never be classified as having potentially long life. In practice, such ideal screening procedures do not exist. The objective of the various procedures described in the following sections is to approach the ideal state as closely as possible. In one procedure the analysis involves calculation of the probabilities of misclassification. Thus one can judge the utility of the procedure on the basis of these probability measures. The remaining simpler procedures do not explicitly involve evaluation of utility; however, theoretically, the analyses could be extended to include such evaluation.

The screening procedures in the following sections are given in order of increasing complexity.

1. Screening by truncation of distribution tails.
2. "Interference" between stress and strength distributions.
3. "Burn-in" screening.
4. Linear discriminant screening.

The description of each procedure includes a discussion of applications, a statement of the assumptions involved, a step-by-step accounting of the procedure, and an illustrative example.

METHODS OF SCREENING

The four screening procedures listed above are paired into two groupings: (1) screening to obtain quality performance at $t = 0$, and (2) screening to obtain long life. The essential difference between the two is that the first grouping involves screening on the basis of initial measurements, whereas the second grouping involves, in addition to the initial measurements, screening on the basis of early life measurements. These may take the form of Δ increments, rates of change in given parameters, etc. The latter type of screening is generally more powerful, but is also more costly.

SCREENING BY TRUNCATION OF DISTRIBUTION TAILS

The primary purpose of this type of screening is to attain device homogeneity in terms of important device parameters with respect to a design specification. Thus the screening problem consists of specifying the device parameters to be controlled and establishing tolerance limits defining the acceptable range of measured values for each parameter. Screening is then accomplished by eliminating those devices from a set having values outside the given tolerance limits. In concept, if the initial parameter values for a population of a given device are described by a probability distribution function, this procedure (as shown) consists of setting tolerance limits on the distribution function so that the distribution tails are truncated (Figure 1).

As shown in Figure 1, the tolerance limits are characteristically expressed in terms of some acceptable percent deviation from the nominal value of the quality parameter. For example, the design specification for a given equipment may specify that all resistors of a given type will have initial resistance measurements within 0.5% of nominal. Thus the tolerance limits for a 10 K ohm resistor would be T_L = 10K − (0.005) (10 K) = 9.95 K and correspondingly T_U − 10.05 K. For devices having several quality parameters, this procedure can be repeated for each parameter, yielding a set after screening consisting of those devices in the original set having all initial parameter measurements within their respective tolerance limits.

From the standpoint of reliability, the concept involved in this type of screening is characteristic of the marginal tolerance limit problem. That is, it is tacitly assumed that devices whose initial parameter measurements are within marginal tolerance limits (T_L, T_U) are not likely to fail in a specified application. Failure is said to occur if a quality parameter exceeds failure tolerance limits (F_L, F_U) (Figure 2).

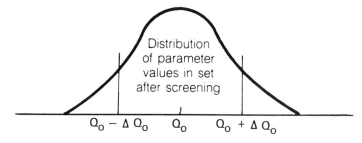

Figure 1 Truncation of the tails of distribution of quality parameter values for a population of devices.

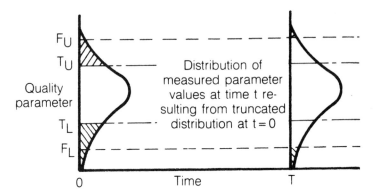

Figure 2 Device reliability at time T resulting from screening out devices out-side marginal tolerance limits at $t = 0$.

This figure shows the distribution of parameter values at $t = 0$. If the devices that have parameter values outside the marginal tolerance limits are eliminated, the resulting truncated distribution will take on the form shown at time T. The proportion of values inside the failure tolerance limits represents device reliability attained as a result of the screening procedure.

The device itself may not fail per se, but its output values exceeding F_L of F_U may induce failure in some other component in the system. The importance of this point is that device tolerance limits must be established with respect to the input-output functions of the device in its intended application. Hence, appropriate tolerance limits should be established based on an analysis of the dynamics of the system of which the devices being screened are a part.

Step-by-Step Procedure

From the foregoing discussion, screening by truncation of distribution tails consists of four sequential steps:

1. Define the set of quality parameters on which basis a set of devices is to be screened.
2. Determine tolerance limits for each quality parameter based on required device input-output functions for system operation.
3. Obtain initial parameter measurements for each device in the set.
4. Screen out the devices that have one or more initial parameter values outside specified tolerance limits.

Thus, screening by truncation of distribution tails has advantages in that it is simple to apply and depends only on parameter measurements at $t = 0$. On the other hand, this procedure has disadvantages in that device reliability parameter tolerance limits require an analysis of the system dynamics which may be very complex.

As an example, consider the problem of selecting reference diodes for application in power supply units. Circuit analysis shows that the output voltage of the power supply (the system) is very sensitive to the stability of the reference diode.

Step 1. The diode parameters, which must be controlled to attain stability, are reference voltage and zener impedance.

Step 2. Tolerance limits for these parameters are detrimental at a specified current level according to the functional relation of diode stability and power supply output. For reference voltage, a two-sided tolerance limit is determined of the form $V_0 \pm \Delta V$, and for zener impedance, a one-sided limit is determined of the form $R - R_0$ (zener impedance essentially represents the slope of the $i = V$ curve at the reference current level). The greater the slope, the less sensitive the reference voltage is to fluctuations in current.

Step 3. Given these tolerance limits, initial measurements are generated for diodes to be used in the power supply unit.

Step 4. The diodes are then screened, based on the initial measurements and parameter tolerance limits.

INTERFERENCE BETWEEN STRESS AND STRENGTH DISTRIBUTION

The purpose of this type of reliability screening is to eliminate devices having strength measurements that may be expected to be exceeded by stress levels in their intended application. The essential difference between this procedure and that given in the preceding section is that here, in addition to obtaining a distribution of device capabilities, it is necessary to generate a distribution of environmental stresses. Reliability is then given by the probability of stress exceeding device strength. Therefore, if those devices having initially low strength measurements are screened out, reliability will presumably be increased.

Two significant problems arise in applying stress-strength analysis: (1) the distribution of environmental stresses is difficult to obtain, (2) the strength distribution will normally change as a function of time.

The environmental stresses will be combinations of many stress components such as temperature, vibrations, pressure, etc. Moreover, the effect on a device of the stress components will not be additive, but will depend on a function of the interactions of the stress components.

Also, the strength distribution will change over time. The rate and type of change will be a function of the environmental stresses actually operating on the device.

In view of the above discussion, the screening problem is therefore to determine an initial value of device strength s^*, so that all devices having measured values less than s^* will depend on the expected environmental stresses in application and on the expected change of the strength distribution as a function of stress.

Mathematical Development of the Screening Procedure

Consider the distribution of environmental stress and device strength shown in Figure 3. The probability that a device has strength s_0 or greater is seen from Figure 3a to be

$$P\{s > s_0\} = \int_{s_0}^{\infty} C(s)\,ds \tag{1}$$

where $C(s)$ = probability density function of component strength.

The probability of a stress value occurring in an infinitesimal interval ds about the point S_0 is given by

$$P\left\{\left[s_0 - \frac{ds}{2}\right] \leq s \geq \left[s_0 + \frac{ds}{2}\right]\right\} = E(s)\,ds \tag{2}$$

where $E(s)$ = probability density function of environmental stress. Therefore the condition reliability given a stress in the interval $[s_0 \pm ds/2]$ is the product of Eq. (1) and (2).

$$dR = \left[\int_{s_0}^{\infty} C(s)\,ds\right] E(s)\,ds \tag{3}$$

Component reliability is obtained from Eq. (3) by integrating this equation over all possible values of s,

$$R = \int_0^{\infty} \left[\int_s^{\infty} C(s)\,ds\right] E(s)\,ds \tag{4}$$

Equivalently, reliability can be determined by considering the probability of a stress less than or equal to s_0 and a component strength of the interval $[s_0 \pm ds/2]$. This leads to the equivalent equation

$$R = \int_0^{\infty} \left[\int_0^s E(s)\,ds\right] C(s)\,ds \tag{5}$$

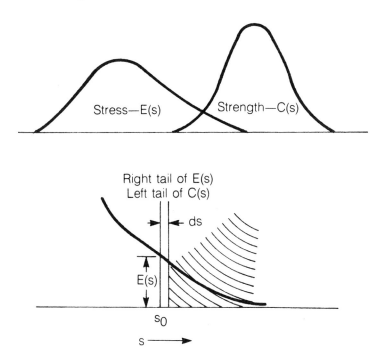

Figure 3 Schematic illustrating reliability as a function of interference between stress and strength distributions. (a) Interference of stress and strength distributions, $E(s)$ and $C(s)$. (b) Graphical description of the calculation of reliability as a function of the interference of $E(s)$ and $C(s)$.

Analytic solution of these equations for reliability is not generally possible; therefore, graphic techniques must be used. If $C(s)$ and $E(s)$ are normal density functions, R can be evaluated by considering the variable $u = (s_2 - s_1)$, where s_2 denotes component strength and s_1 denotes environmental stress. Clearly, no failure will occur if $u = (s_2 - s_1) \geq 0$. Since s_2 and s_1 are normally distributed, $u = s_2 = s_1$ is also normally distributed with mean

$$\mu(u) = \mu(s_2) - \mu(s_1) \tag{6}$$

and variance

$$\sigma_u^2 = \sigma_{s_1}^2 + \sigma_{s_2}^2 \tag{7}$$

where $z = (u - \bar{u})/\sigma_u$. The integral of Eq. (9) can be readily evaluated by using the table of the standard normal distribution. In Figure 4, reliability is plotted as a function $\bar{u}/\sigma_u$, the separation between the means

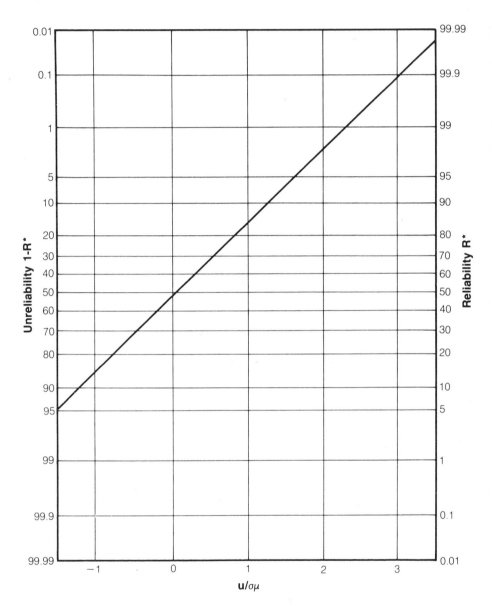

Figure 4 Component reliability as a function of the distance, u/u, between the mean values of the stress and strength distribution E(S) and C(S).

of $C(s)$ and $E(s)$ measured in units of the standard deviation σ_u. For example, if $\bar{u}/\sigma_u = 1.25$ component reliability is seen to be $R = 0.90$.

$$R = P\{u > 0\} = \frac{1}{\sqrt{2\pi}\sigma_u} \int_0^\infty \exp\left[\frac{u - \bar{u}}{2\sigma_u}\right]^2 du \qquad (8)$$

$$= \frac{1}{\sqrt{2\pi}} \int_{\bar{u}/\sigma_u}^\infty \exp(-z^2/2)\,dz \qquad (9)$$

Alternatively, if an $R = 0.99$ is required, a separation of $\bar{u}/\sigma_u = 2.28$ must exist between the mean of $C(s)$ and $E(s)$.

The Screening Problem

Figure 4 and the foregoing discussion show that reliability increases as the area of the overlap region of $C(s)$ and $E(s)$ decreases. For given distribution function $C(s)$ and $E(s)$, reliability can be increased by screening out those components having strength measurements in some part of the overlap region. The screening problem is, then, to choose a critical value of strength s^* so that all components having strength measurements less than s^* are eliminated. Since the procedure consists of truncating the left tail of the $C(s)$, this type of screening is similar to that described earlier.

For given $C(s)$ and $E(s)$, s^* should be chosen so that reliability resulting from screening will attain a specified requirement, R^*. If we substitute the truncated distribution of strength values, $C_T(s)$, for $C(s)$ in Eq. (4), we obtain

$$n = 2\left[\frac{(t_1 + t_2)s}{d}\right]^2 \qquad (10)$$

$$R^* = \int_0^\infty \left[\int_s^\infty C_T(x)\,dx\right] E(s)\,ds \qquad (11)$$

$$C_T(s) = \begin{cases} C(s)\int_{s^*}^\infty C(s)\,ds, & s \geq s^* \\ 0, & s < s^* \end{cases} \qquad (12)$$

Since $\int_{s^*}^\infty C(s)\,ds$ for given s^* is a constant with respect to the variable of integration, Eq. (10) can be written as

$$R^* = \int_{\infty}^{s^*} E(s)ds + K(s^*)\int_{s^*}^{\infty}\left[\int_{s}^{\infty}C(x)\,dx\right]E(s)ds,$$

$$K(s^*) = \left[\int_{s^*}^{\infty}C(s)ds\right]^{-1} \tag{13}$$

Equation (12) cannot be solved analytically for the value of s^* that will yield a specified reliability R^*; hence, numerical approximation methods must be used. However, if s^* is sufficiently far out in the left tail of $C(s)$, a close approximation to Eq. (12) is

$$R^* = E(s^*)\left[\int_{0}^{\infty}\left(\int_{s}^{\infty}C(x)\,dx\right)E(s)ds\right] \tag{14}$$

The quantity inside the brackets of equation 14 is seen to be identical to R [Eq. (4)]. Thus, we obtain

$$R^* = K(s^*)\,R \tag{15}$$

That is, the reliability obtained by screening based on a given s^* is equal to the reliability before screening multiplied by a constant. This approximation will yield a reasonably small overestimate of reliability for values of s^* in the lower 5% tail of the distribution of $C(s)$.

Based on the above, a suitable procedure for determining s^* so that a specified reliability after screening, R^*, is obtained is as follows:

1. Calculate the required s^* by using the approximating function Eq. (15).
2. Determine whether the s^* calculated in step 1 is sufficiently far out in the left tail of $C(s)$, say $s^* \leq s\alpha$ where $s\alpha$ is the lower percent point of $C(s)$.
3. If $s^* \leq s\alpha$, it can be used as the required critical value for screening. If $s^* > \alpha$, then a new value of s^* must be determined numerically that satisfies the exact function in Eq. (12).

For example, suppose that when $C(s)$ and $E(s)$ are normal distributions, a reliability after screening of $R^* = 0.95$ is required. From Eq. (15), $K(s^*) = R^*/R$ where R depends on $\bar{u}/\sigma_u$ as shown in Eq. (9) and Figure 4. Values of s^* satisfying the requirement $R^* = 0.95$ are given in Table 1 for various values of R.

Table 1 shows, as expected, that as the distance between the stress and strength distributions increases, the size of the tail of the strength distribution that is truncated decreases for a given R^*. With respect to

Table 1 Values of the Critical Values s^* such that $r^* = 0.94$[a]

R	$\bar{u}/\sigma_u$	$K(s^*) = R^*/R$	$1/k(s^*)$	s^*
86	1.05	1.10	0.906	−1.32
88	1.15	1.08	0.927	−1.45
90	1.26	1.05	0.948	−1.63
92	1.37	1.03	0.969	−1.88

[a]For various values of $R(s^*)$ is given in units of the standard deviation σ, of strength distribution $C(s)$.

the acceptability of the approximate values of s^* in Table 1, it is seen that for $R = 0.90$ and 0.95 that s^* is in the lower 5% tail of $C(s)$ and, therefore, will yield a reasonably good approximation. On the other hand, the s^* values for $R = 0.90$ will yield a significant overestimate of R^*. Thus, for the latter cases, it is necessary to resort to numerical determination of s^* satisfying Eq. (12).

To make clear the conceptual problem of choosing s^*, consider the illustration in Figure 5. As s^* is made continuously greater [that is, s^* moves further out in the left tail of $C(s)$], the area of the overlap region of $C_T(s)$ and $E(s)$ will decrease, and therefore reliability will increase. However, the rate of change of R will decrease with increasing s^*. Moreover, as s^* increases, the number of components eliminated by screening will increase at an increasing rate. If we assign some cost to each component screened out, we can reasonably assume that at some point the incremental increase in reliability is not worth the cost incurred in attaining it. The important point is that s^* should be chosen with respect to both reliability requirements and the cost of attaining the required reliability, as it is subjected to the "law of diminishing returns." Note the decreasing amounts of ΔR for a given ΔS as s^* is increased. The maximum point is reached, of course, where the slope of the curve is zero (0).

Step-by-Step Screening Procedure

1. Determine the distribution of strength measurements, $C(s)$, for the set of components to be screened.
2. Define the environmental stress variables acting on a component in its intended application.
3. Determine an equation for combining the stress variables into a single measure, s.

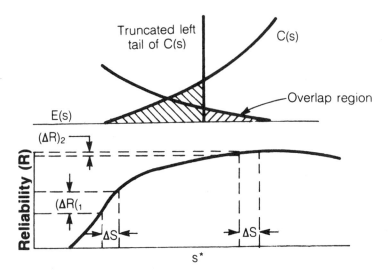

Figure 5 Incremental gains in reliability for a given incremental increase in s^* as a function of s^*.

4. Determine the distribution of the environmental stress measure, $E(s)$.
5. Determine the critical values s^* on the basis of a specified reliability requirement, R^*, from Eq. (12) or (14) and (15).
6. Eliminate the components that have strength measurements less than s^*.

BURN-IN SCREENING

The purpose of burn-in screening is to eliminate components that have inferior reliability on the basis of a short-term environmental test. Environmental tests may be run under high stress conditions or under normal stress conditions with respect to some intended application. Components eliminated as a result of the test may actually fail during the test, or may exhibit performance characteristics indicative of early failure in actual application.

The development of a burn-in screening procedure involves consideration of (1) the assumptions associated with the procedure, (2) the test design, and (3) the screening criterion. The essential aspects of these factors are discussed in the following paragraphs.

Assumptions in Burn-In Screening

The underlying assumption in burn-in screening is that a set of components before screening consists of two subsets: (1) a subset of superior components represented by a distribution of strength measurements with a high mean strength and small variance, and (2) a subset of inferior components represented by a distribution of relatively low strength measurements (Figure 6).

In terms of failure rate, this assumption (illustrated in Figure 6) implies a failure rate curve of the familiar bathtub form shown in Figure 7. Inferior components fail early, yielding an initial high failure rate. After this initial period (infant mortality), the failure rate curve levels off until the superior component strengths degrade to the point of wearout.

A second assumption concerns the functional relation of parameters on which screening is based to the failure mechanisms acting in a component. If every failure mechanism and its physical properties for a given component were known, it would be possible to determine what parameter measurements act as precursors of component failure. However, all the failure mechanisms are not generally known or are not completely describable. Therefore, we must assume (to some degree) which parameters should be measured as precursors of failure.

A third assumption, closely related to the one above, concerns characteristics of parameter variation that indicates early failure. These characteristics may include parameter instability, parameter drift in some time increment, or rate of parameter drift. The degree of assumption in this case will depend on the knowledge of the degradation processes acting in a given component type.

The first assumption above is statistical, whereas the second and third assumptions are physical in nature. The significant point is that as our physical knowledge of component features increases, the degree of uncertainty in the physical assumptions decreases. This implies that if we gain complete knowledge of the physical degradation processes acting in a component, we can predict early failures with near certainty on the basis of short-term burn-in tests.

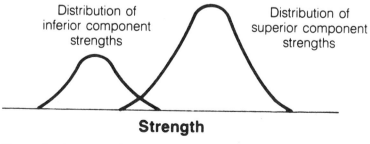

Strength

Figure 6 Assumed distribution of component strengths before screening.

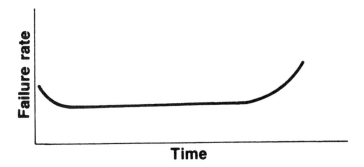

Figure 7 Schematic representation of assumed strength distribution in terms of failure rate.

The Environmental Test

The test design for burn-in screening requires specification of (1) the parameters to be measured, (2) the stress levels for each of the stress factors, and (3) the time duration of the test.

Parameters to be measured in a particular instance depend on the degradation processes acting in the component and how these processes relate to the total set of measurable parameters. The parameters will be specified on the basis of subjective experience, as discussed in the preceding section on assumptions. Once the parameters have been specified, the next step is to determine what measure of the parameter should be used as a basis for screening. As discussed previously, these may include such measures as instability, incremental change over some period, and rate of change.

A second factor to be specified in the test design is the stress levels under which the test should be run. The governing principle here is to operate the components under a stress environment for a specified time such that the inferior components will fail or exhibit undesirable parameter variation while the superior parts are not prematurely aged. Thus the problem is to select stress levels that are sufficiently severe to yield significant parameter degradation in the inferior components without introducing failure mechanisms not existing under normal stress conditions.

The third factor in the test design is the time duration of the test. Ideally, the test time should be that time required to yield statistically significant parameter degradation at the highest possible stress levels giving "true" acceleration. In practice, however, the appropriate trade-off between test duration and stress level must be determined largely from experience of previous test analyses. Again, as knowledge of the reliability physics increases, greater precision can be introduced into the screening test design. Table 2 lists stress levels and test times that characteristically used for various component types in burn-in tests. Notice that the stress conditions given in this table are intended to be illustrative of stresses that actually may be applied. Alternative stress schemes may be applicable, such as increasing stress testing, short-term overstress, etc.

The Screening Criteria

The final step in developing a burn-in screening program is to determine screening criteria. These criteria generally are in the form of tolerance limits on the measured parameters. The concept in determining these limits is the same as that used in screening by truncation of distribution tails, previously discussed. That is, the limits can be considered as marginal tolerance limits in that, if a component parameter is outside of marginal tolerance limits after the burn-in test, the probability is high that the parameter will be outside of failure tolerance limits at some early time of actual operation. These relations can be estimated from long-term component life tests.

Step-by-Step Procedure for Burn-In Screening

1. Identify the failure mechanisms for the component type to be screened.
2. Determine the measurable parameters that are indicative of the failure mechanisms.
3. Specify the environmental stress levels and time duration for the burn-in test.
4. Specify the desired measure of the indicative parameters, environmental change, rate of change, etc.
5. Establish the screening criteria: tolerance limits on the measured parameters.
6. Run the test and screen out the components that have parameter measures outside of tolerance limits.

LINEAR DISCRIMINANT AS A BASIS FOR SCREENING

Linear discriminant, as a basis for screening components, may be considered as a logical extension of the techniques of the screening procedures given in the preceding sections. Its objective is to predict early failures on the basis of a weighted average of initial and/or early life parameter measurements. This weighted average is called a linear discriminant.

Table 2 Stress Levels and Test Times Commonly Used in Burn-in Test Design.

Component	Test Time	Stress Levels
Transistors	125-250 hours	Near rated power dissipation. Temperature approximately 90% of maximum operating temperature.
Diodes	500 hours	Near rated load. Temperature approximately at maximum operating temperature.
Resistors	750 hours	Electrical stress approximately 80% of rated load. Temperatures in excess of rated temperature.
Capacitors		Electrical stress at rated voltage. Temperature at rated value.
Microelectronics	160 hours	at 125°C

The problems in constructing a linear discriminant are to determine which parameters are significant in predicting failure, to determine the "weight" that should be assigned to those parameters, and to determine what limits on the value of the linear discriminant should be used in order to reject the inferior (or select the superior) components.

In order to construct the linear discriminant, it is necessary to conduct a load-life test on a statistically significant number of the components that are to be screened. Results of this life test are used to define failure (if this has not been previously dictated by the application), to select the parameters that will be used in the linear discriminant, and to evaluate the probability of success of the discriminant in screening subsequent lots of the component parts. Thus, in contrast to the procedures previously discussed, it is seen here that the parameters on which screening is based and the predictive "power" of the screening function are determined as explicit steps in the analysis.

Screening is carried out by measuring a set of p different parameters selected in the indiscriminant analysis for each component. These measured values $x_1, \ldots, x_p$, are then substituted into an equation of the form: $z = \lambda_1 x_1 + \cdots + \lambda_p x_p$, where the λ's denote optimal numerical "weights" associated with the measured values. This equation is called a linear discriminant. If the weighted average, or Z value, is less than a critical value of Z, the component is judged to be a potential early failure. Otherwise, the component is judged to be a potential late failure, or in other words, to have a long life.

An ideal screening criterion would never misclassify a component. That is, a superior component would never be classified as inferior and, conversely, an inferior component would never be classified as superior. In practice, such ideal screening criteria do not exist. However, the use of linear discriminant analysis permits the construction of screening criteria for which the probability of misclassification can be computed when certain assumptions are made. Consequently, if the probabilities of correct classification are small, the screening criterion is of little practical significance, even though the criterion may show statistical significance.

The principal advantage of using the method of linear discriminant analysis to construct the screening criterion is the fact that no other linear combination of parameter measurements will yield a screening criterion with small probabilities of misclassification. Thus, if the linear discriminant method fails to produce a screening criterion of practical significance, the search for a linear screening criterion may be abandoned. This remark is qualified to the extent that certain assumptions must be made and that various transformations of the data may also be required. If these fail, still other modifications can be attempted in an effort to improve the screening classification.

Development of a Linear Discriminant

The development of a linear discriminant screening criterion requires the following steps:

1. A load-life test must be performed on a statistically significant number of components. Components selected for the test must be representative of the population of components for which the screening criterion is being derived. The length of the test is that time for which reliability prediction is desired.
2. The linear discriminant screening criterion is then to be derived from initial, early life, and end-of-test parameter measurements of the components on test.
3. The resultant screening criterion should be empirically verified prior to screening components of unknown reliability.

These steps are described in detail in the following paragraphs along with assumptions associated with this screening procedure.

Assumptions Involved in Linear Discriminant Analysis

Linear discriminant analysis is essentially a form of linear regression analysis. As such, it involves the same assumptions that are pertinent to statistical regression analysis. In particular linear discriminant analysis requires that:

1. The parameter values used in the linear discriminant function must be normally distributed.
2. The within-sets variances must be homogenous. That is, the variance of the parameter measurements of the superior components must be the same as that for the inferior components.

Where these assumptions do not hold true, various transformations may be tried in order to satisfy the assumptions. For example, it is frequently the case that several of the measured parameters are decidedly not normally distributed. It has been noticed, in many of these instances, that a logarithmic transformation of the parameter measurements yields satisfactory agreement with the normality assumption.

The Experimental Procedure

The first step in deriving a linear discriminant screening criterion consists of a load-life test on a sample of components drawn from the population of the component type of interest. This experiment consists of the following steps:

1. Define component failure. Failure may be defined either in attribute form (operative or inoperative) or in variables form (tolerance limits on one or more parameters).
2. Select the time period, t, for which reliability prediction is to be made. Thus, t represents the desired life of the component in its intended application.
3. Select a sample of components representative of the population for which the screening criterion is being derived. Appropriate statistical randomization procedures should be used in selecting the sample.
4. Determine the parameters to be used as potential predictor of failure. These may be determined from engineering judgment and/or prior test data.
5. Obtain initial parameter measurements. If parameter change over a short time period, Δt, is considered as a potential predictor, then parameter measurements must be obtained after the operating interval Δt.
6. Run the life test for a time period, t, under actual or simulated operating conditions to be encountered in the intended application of the component.

7. At the end of the test, obtain the parameter measurements in the same terms as failure is defined for each component.
8. Separate the components into two sets: S_1, consisting of those components that did not fail during the test, and S_2, consisting of those components that failed.

As an example of a criterion of failure, in a twofold classification, each component may be judged to have failed, or not to have failed, at the end of the test. The criterion of failure may be based on an attribute definition: those components may be called failures which are not operative at the end of the test. All others are judged not to have failed. Alternatively, it may be that all of the components are operative at the end of test, those component parts being called failures which exhibit a drift in excess of some specified value of some measured parameter. The importance of the failure definition stems from the fact that the screening criterion, determined by the linear discriminant analysis, attempts to reproduce from initial or early life measurements that same classification assigned at the end of the test. The prediction classification may be based on initial measurements taken at $t = 0$ or on early life measurements taken, for example, at $t = 0$ and $t = 100$ hr.

Linear Discriminant Analysis

For each set of measured parameters, there exists a linear discriminant screening criterion. For example, if three parameters are measured, it would be possible to construct seven different linear discriminants. Three of these would involve each parameter, singly; three would involve a combination of two parameters; and one would involve all three parameters. By examination of the probabilities of misclassification associated with each of these screening criteria, it is possible (1) to determine the minimum number of parameters required for practical screening; (2) to determine the gain in reliability obtained by screening on the basis of two parameters rather than one, three rather than two, etc; and (3) to determine the parameters that may be interchanged for measurement or cost reasons without changing the effectiveness of the screening procedure. When many different parameter measurements are available, it is not feasible to examine all possible subsets and their associated linear discriminants.[1] Various methods may be used to select the parameters that appear to be the most suitable for inclusion in a linear discriminant. (See for instance our discussion on factorial designs.)

The steps below describe the computations required for obtaining a linear discriminant screening criterion, given the parameters to be used as predictors.

   Construction of the reliability screening criterion, based on a specified set of parameters, consists of seeking a quantitative measure of the
difference between the superior and the inferior components. This
measure should be based on initial and early-life measurements and
should serve to predict the qualities of the component during late life.
Let S_1 denote the set of initial and early-life measurements associated
with the superior components and S_2 the measurements associated with
the inferior components. We want to define a "discriminating" function
of the measurements in S_1, which assumes values that are distinctly different from the values assumed by the function for the measurements in
S_2. An ideal discriminating function would be one that yields 1 for all
measurements associated with superior components and yields 0 for
measurements associated with inferior components. In practice, ideal
discriminating functions do not exist. Instead, an attractive approximation to such a function is given by a linear combination of the measured
parameters

$$Z = \lambda_1 \chi_1 + \cdots + \lambda_p \chi_p \qquad (16)$$

where χ_i, $i = 1, \ldots, p$, denotes a measured value of the ith parameter,
and the λ's are numerical "weights." When the optimal λ's are determined as described below, the linear combination of measured values is
called a "linear discriminant." The optimal λ's are those values of the λ_i
that maximize the quantity $(\bar{z}_2 - \bar{z}_1)/s$. That is, the optimal λ's maximize the distance between the means of the distributions of superior and
inferior components, $\bar{z}_1$ and $\bar{z}_2$, respectively, expressed in units of the
standard deviations. The statistical properties of linear discriminants
were first developed by R. A. Fisher. Mathematical details of such
analyses are given in several previously referenced statistical texts.
Below, the essential computational steps in deriving a linear discriminant are described.

1. Partition the initial and/or early-life measurements of parameters
    $x_{ip}, \ldots, x_p$ into sets: (a) those associated with components S^1, and
    (b) those associated with components in S^2.
2. Compute the sums of squares and products, within the set of supe-
    rior components, S^1_{pq}. Repeat for the inferior components, S^2_{pq}.
    These are given by

$$S^1_{pq} = \Sigma \, i \, \Sigma_j \, (\chi^1_{ip} - \chi^{-1}_p)(\chi^1_{jp} - \chi^{-1}_q) \qquad (17)$$

where $\chi^1_{ip} =$ the value of parameter p of the ith component in S^1
and $\bar{\chi}^1_p =$ the average value of parameter p in S^2. S^2_{pq} is obtained
in the same way.

$$S_{pq} = S^1_{pq} + S^2_{pq} \qquad (18)$$

3. Solve Eq. (19) for λ, which gives the λ-weights for the linear discriminant [Eq. (15)].
4. Compute the quantity

$$S\lambda = d \tag{19}$$

$$S = \begin{bmatrix} S_{11} & \cdots & S_{1p} \\ \cdot & & \cdot \\ \cdot & & \cdot \\ \cdot & & \cdot \\ S_{p1} & \cdots & S_{pp} \end{bmatrix}, \quad \lambda = \begin{bmatrix} \lambda_1 \\ \cdot \\ \cdot \\ \cdot \\ \lambda_p \end{bmatrix}, \quad d = \begin{bmatrix} \varkappa_1^{-1} - \varkappa_1^{-1} \\ \cdot \\ \cdot \\ \cdot \\ \varkappa_p^{-1} - \varkappa_p^{-2} \end{bmatrix}$$

$$\delta = \frac{\bar{z}_2 - \bar{z}_1}{s} = \left[(n_1 + n_2 - p - 1) \sum_{i=1}^{p} \lambda_1 d_i \right]^{1/2} \tag{20}$$

where n_1 is the number of components in S^1 and n_2 the number in S^2. The quantity $\varkappa$ denotes the distance between the mean values for superior and inferior components in units of the standard deviation s. $\varkappa$ constitutes a measure of the power of the screening criterion in that the greater $\varkappa$, the smaller the probabilities of misclassification.

5. Compute the probabilities of classification as shown in Table 11.3. For example, the probability of disclassifying a "truly" inferior component as superior is seen to be $1 - \phi(\delta/2)$.
6. Repeat steps 1 to 7 for each combination of parameters of interest and compare the respective "powers" (δ) and resultant probabilities of classification. Acceptance of a given discriminant function will depend on its relative power balanced against the "costs" of application.

Because of the assumptions that must be made in the course of developing the criterion and because these assumptions are often incapable of direct verification, it is desirable to make an indirect check on the validity of the linear discriminant screening procedure. This is accomplished by applying the linear discriminant screening criterion to screen the components used to generate the test data. That is, the initial and early life measurements are substituted into the discriminant, and each component part is classified by comparing its z value with the critical value of z. Notice that this is principally an internal consistency check and, as such, does not validate the general applicability of the linear discriminant.

Ideally, the linear discriminant will produce the same classification assigned to the component parts at the end of the test. The extent to which the linear discriminant achieves the end-of-test classification is an indication of the practical effectiveness of the discriminant screening criterion. If the end-of-test classification is not closely approximated by the discriminant screening based on early-life measurements, the screening criterion is judged of little practical significance. This result may indicate that none of the measured parameters is relevant to predicting the correct classification. Alternatively, it may indicate that some assumption is incorrect and that transformations of the data should be considered as a method to improve the classification.

If the linear discriminant yields good agreement with the classification at the end of the test, it may be applied to other sets of component parts of known quality that were not used in its construction. If the discriminant passes all verification checks successfully, it may be used to screen component parts of unknown quality that have similar statistical properties. The more thorough the verification the more assurance we have that the screening procedure will perform properly.

Application of the Linear Discriminant Screening Criterion

Once a linear discriminant has been developed, sets of component parts of unknown reliability can be screened by

1. Determining a critical value, z^*
2. Obtaining a z value for each component to be screened
3. Classifying as superior the components for which $z < z^*$, and classifying as inferior the components for which $z > z^*$ (this assumes that $\bar{z}_2 > \bar{z}_1$)

The expressions for the classification probabilities given in Figure 8 assume that z^* is midway between z_1 and z_2. In general, however, z^* may be chosen to satisfy some specified criterion function. Such criteria may include: (1) maximization of cost of misclassification, or (2) specifi-

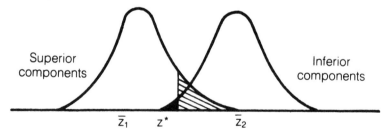

Figure 8   Schematic representation of the theoretical distribution of z-values for the superior and inferior components.

Table 4 Expected Conditional Probabilities of Classification[a] [Given $z^* = (a\bar{z}_1 + (1-a)\bar{z}_2)$]

Assigned Classification	Conditional Probability of Classification with Linear Discriminant Criterion	
	Superior	Inferior
Superior	$\phi[(1-a)\delta]$	$1 - \phi[(1-a)\delta]$
Inferior	$1 - \phi(a\delta)$	$\phi(a\delta)$

[a] $\phi(y) =$ the area to the left of y under the standard normal distribution curve.

cation of some acceptably small probability of misclassifying an inferior component as superior. Thus if we defined the critical value by

$$z^* = a\bar{z}_1 + (1-a)\bar{z}_2, \quad 0 \le a \le 1 \tag{21}$$

then the expected conditional probabilities of classification are as given in Table 4.

Relation of Linear Discriminant Analysis and Physical Degradation Laws

Parameters characteristically used in a linear discriminant screening criterion are operational parameters. That is, they are generally the parameters in terms of which component performance is measured. For example, a criterion for screening transistors may be based on initial or early life parameter measurements such as collector cutoff current, emitter cutoff current, collector breakdown voltage, etc. How well this class of variables serves as predictors of component reliability is determined from statistical analysis of test data as opposed to theoretical considerations. The net effect of reliability prediction based on this class of variables is that their potential "power" is bounded by the degree of statistical variation in the operation parameters serving as measures of the fundamental aging process acting in a component.

If we can uncover those fundamental parameters characterizing aging processes such as activation energies, frequency factors, diffusion coefficients, etc., then significantly more powerful screening criteria can be developed. Moreover, the screening criteria based on this more fundamental class of parameters may take on or be closely related to the form of the physical laws governing aging processes.

For example, suppose that activation energy and frequency factor were taken as the the parameters for a linear discriminant screening criterion. The analysis may show that the linear discriminant

$$a = \ln(\text{frequency factor}) = \frac{1}{kt}(\text{activation energy}) \qquad (22)$$

should take on the form

$$= \ln A = \frac{1}{kt}\Delta E \qquad (23)$$

where $\lambda_1 = 1$, the coefficient of the pre-exponential term, and $\lambda_2 = -1/kt$. From Eq. (23), it is seen that the rate of aging R is related to the discriminant z-value by

$$R = e^z = Ae^{-\Delta E/kt} \qquad (24)$$

Thus, in terms of linear discriminant screening, this example states that the optimal discriminant is dependent on the logarithmic aging rate of a component. In other words, the discriminant that maximizes the distance between the distributions of z-values for superior and inferior components depends on only a single parameter: the aging rate.

An important result of deriving linear discriminants from more fundamental parameters directly related to aging process acting in components is that several significant problems in executing a linear discriminant analysis may be alleviated. These problems include the following ones:

1. Physical theory may preclude the necessity of testing many parameter combinations to find the one that is optimal. Only a small number of tests may be required, each involving a small number of parameters. Thus the computational problem will be greatly reduced.
2. Physical theory may indicate the type of parameter transformation that should be made, thus yielding the best agreement between expected and observed screening results.
3. Physical theory may provide a basis for verification of the λ weights in the linear discriminant.
4. Physical theory may permit the derivation of linear discriminants under accelerated test conditions that can be used to screen components that have intended application in various other stress environments.

In general, the potential advantage to component screening in reliability physics is that screening criteria based on a more fundamental class of variables can be derived more sharply, can be easier to validate, and may be significantly more powerful.

SCREENING PROCESS

The screening process is normally included in the item specifications. This process, as defined in the specifications, includes tests to be performed and the methods to be followed. For example, a typical capacitor specification would include a tabulation of the tests to be performed similar to those in Table 5.

In addition, the test method to be followed would also be indicated in a manner similar to the following.

Humidity Screening

Capacitors must be subjected to a 24-hour 100% relative humidity test at 85°C. Upon completion of the test, the capacitors should be allowed to stabilize at room ambient conditions for a period of one hour. After stabilization, capacitance, insulation resistance, and dielectric withstanding voltage must be measured and recorded as required by the item specification.

Voltage Stabilization

Capacitors must be subjected to a test at a predetermined ambient temperature. Upon completion of the test, capacitors should be allowed to stabilize at room ambient conditions for a period of ½ hour. After stabilization, the capacitance, dissipation factor, and DC leakage should be measured and recorded.

Table 5   Screening Process, 100% Inspection

Test
1. Humidity screening
2. Seal
3. Voltage stabilization
4. Intermittent high temperature operation
5. Temperature cycling
6. Dielectric withstand voltage
7. X-ray
8. Visual inspection
9. Temperature coefficient and capacitor drift
10. Capacitance
11. Dissipation factor (quality factor)
12. Insulation Resistance
13. DC leakage

Intermittent High Temperature Operation

Capacitors must be subjected to an intermittent operation test at their rated maximum ambient temperature (for full-rated voltage). Test duration should be 8 hours ± 15 minutes. Direct-current voltage must be applied to the capacitors according to the specification. The surge current must be limited to 50 mA. Upon completion of the test, the capacitors should be stabilized at room ambient conditions for a period of 16 to 24 hours. After stabilization, capacitance and dissipation factor must be measured and recorded. Capacitors containing liquid impregnant should be examined under ultraviolet light (black light) for evidence of impregnant leakage.

SUMMARY

Screening tests offer the most meaningful technique whereby products requiring high reliability performance can be supplied. Provided that a quality-oriented manufacturing cycle is followed, a manufacturer can segregate his output and grade it according to preestablished reliability standards. Although all products will not comply with the highest requirements, the manufacturer's reject rate is not prohibitive, since he can supply his output to varying degrees of customer reliability and quality requirements. The most common techniques employed for this purpose have been presented in this chapter. Lest the reader think that what has been presented in this Chapter is all there is on this subject we have included the following recommended readings. These references include all that has been said above and deserve the attention of anyone intending to implement these techniques.

RECOMMENDED READING

Proceedings Annual Symposium on Reliability 1954–84. New York: The Institute of Electrical and Electronics Engineers, Inc.

NOTE

[1]The problem in this case is that such discriminant derivation requires the inversion of a matrix, which is an expensive process. Obviously, if a sufficiently large computer is available, greater numbers of parameter combinations can be tried.

Chapter 12

Software Evaluation
Philosophy and Approach

Products increasingly incorporate computer programs as an integral part of their design. In addition, software design and production is a major industry in and of itself. Therefore, software evaluation through the design and use of effective test programs plays an important role in software quality and reliability. This chapter treats software testing from a philosophical and practical point of view. Details of software testing and evaluation are beyond the scope of this limited treatment however.

SOFTWARE DEVELOPMENT CYCLE

There are several critical activities in the development and evaluation of high-quality software. Visibility into and control of these activities are essential to both the software developer and the user.

To avoid the common occurrence where software problems are found later in development and not in the particular phase in which they were created, a systematic and comprehensive design and test approach is needed. Figure 1 depicts a typical development and test effort. Normally the majority of errors are introduced during development, and the majority of errors are discovered and corrected after the product components have been fabricated.

What is needed is an evaluation program designed to demonstrate the quality characteristics of the product defined for evaluation at a particular stage of development. The test plan should provide for independent assessment of the product as it is being developed. Current practice of focusing all evaluation during the testing activity phase is insufficient to produce the desired quality in the software. However, since the focus here is on product test, we now turn to to a discussion of the testing activity phase of software development. The design and use phases are deserving of careful examination as well.

EVALUATION PLANNING

The planning stage of software development needs to receive the same degree of attention as hardware design activities. It is important to emphasize that the software testing goal differs markedly from that of hardware testing. While the assumption of hardware testing is to prove the product is safe, accurate, reliable, and conforms to functional specifications without error, software testing and its planning must proceed from exactly the opposite premise. *The purpose of software testing is to identify as many errors as possible.* As a result, the test planning must proceed on an entirely different philosophical base and one which runs counter to well-established product test and evaluation practices. Failure to adopt this approach from the outset is bound to lead to the underestimation of required resources and in view of scheduling constraints, the avoidance of undertaking the most demanding test program possible. The end result is that errors in the software will be overlooked.

As in the case with hardware testing, the evaluation plan is a vital component of the test process. The elements of a good software test plan are essentially the same as those for a hardware program. However, the nature of software evaluation is significantly different in several aspects. They include:

Test Tools. A wide variety of tools are required for proper testing. They need description and how and when they will be obtained.

Test Case Standards. The key to effective testing lies in the degree of care taken in the preparation of tests. Preparation guidance, storage, and maintenance must be detailed in order to insure effective evaluation.

Success Criteria. The basis for determining when the software has satisfactorily met pre-established performance standards and the successful completion of the test program has been achieved must be determined.

Of the terms listed above, success criteria is the most difficult to establish. The problem is to encourage the detection of as many errors as possible. This in essence involves the establishment of *negative* criteria. Testing is not complete until x number of errors *are* detected.

Anyone familiar with conventional product testing and evaluation can easily appreciate the conceptual gap that must be bridged in the planning stage if a software evaluation plan is to meet its objectives. The idea that the aim of testing is to find errors is fundamental to software testing. Success is defined in terms of the number of errors that are detected. In fact, the more errors one detects and the ease with which they are detected, the more successful one is and the more likely additional errors exist!

A key document is the evaluation plan. An independent evaluation of the plan can help to strengthen a weak plan by identifying the weaknesses and recommending improvements prior to development of the detailed procedure and subsequent implementation. The plan should emphasize the independent evaluation of test results at each level of software testing. The levels referred to include acceptance, installation, and system testing.

Acceptance Testing

Acceptance testing is a validation process that tests the software to the initial requirements. The major characteristic of this testing is that initial requirements are the criteria for evaluating the test results. Special attention is given to data base structures, ranges of values of parameters, and exercising or invoking each instruction and branch statement in the code. Test cases are designed based on the design architecture, processing flows defined in the design, and details defined in design specifications which altogether reflect the requirements criteria. Requirements lists may be used to measure progress in validation testing. Such lists, when developed and applied at detailed level, can be used to decompose a complex, voluminous collection of requirements into checklists to be used by analysts and managers to guide and evaluate the test process. Not only do nominal processing paths need to be emphasized, but so do the unexpected or low probability paths associated with contingency processing. Only through this systematic process will the user organization be able to show how the product does *not* meet the requirements. When all of their tests have "failed," only then does the user know the product is acceptable. Acceptance tests may include a combination of test cases, pilot runs, and on-line testing.

Installation Testing

As with all other types of testing, the objective of installation testing is to find errors. Installation testing is concerned with the errors that may occur during the installation process. This process includes option selection, file and library allocation, and integration with existing programs.

Installation tests should form an integral part of the software and its documentation prepared by the software developer and shipped with the product.

System Testing

The purpose of this type of testing is to find errors between the system and its original objectives. Components of the test include the system itself, the product objectives, and all documentation that will be supplied with the system.

To the extent the system is tested in the end use environment or a simulation, the process is one of validation. However, frequently testing in the actual or simulated environment is not possible and then the system testing is one of verification.

The key to the efficiency of system testing, whether validation or verification, is the design of the test cases. Due to the complexity and scope of system testing, test design requires much ingenuity. The following 14 categories adopted from Myers[1] identify the kinds of tests worthy of consideration.

1. Load/stress testing. The purpose here is to expose the system to peak load conditions to see how it behaves under extreme conditions. While these conditions may occur only infrequently, they do occur and the system should be required to perform under the conditions posed by them.

2. Volume testing. The interest here is to show that the system or program cannot handle the amounts of data specified in the objectives. Voluminous data must be fed into the system over an extended period of time to test its ability to cope with the load placed upon it. An example would be a process control system which indicated simultaneous out-of-control conditions from each control point.

3. Configuration testing. A system is likely to consist of a large variety of hardware and software configurations. The combinations may be extensive. The test program should test the limits, that is the minimum and maximum configurations possible. All hardware and software involved should be included in the test.

4. Compatibility testing. System improvements must be tested for incompatibility with existing and continuing systems. Compatibility objectives should be stated and tests designed to see that they have been met.

5. Security testing. Due to the sensational breeches of system security of late, security objectives are of great importance. The test objective is to compromise the system by gaining unauthorized

access. According to the adage "It takes a thief to know a thief," some test designs provide for an independent expert to subvert the system. The test will be a success if system security is violated.

6. Storage testing. Tests should be conducted to show that the system does not achieve the main and secondary objectives specified in the specifications.

7. Performance testing. Performance and efficiency criteria such as processing time and output lineage, require proving in the various system configurations that are possible. Worst-case configurations should be specified to give the most critical evaluation.

8. Installation testing. The procedure for installation needs to be tested to insure that customers or contractors can install the system within time and cost expectations. Some systems are beyond customer technical or financial capability.

9. Reliability/availability testing. The objective is to demonstrate that the system cannot meet stated reliability and/or availability objectives. These tests are difficult and expensive to conduct but nevertheless efforts should be made to test the objectives of mean-time-to-failure, error rate, and the detection, correction and/or tolerance of both software and hardware errors.

10. Recovery testing. The recovery of function, whether it be from losing data, power failure, or telecommunications failure, requires testing. The system should be exposed to the faults and its recovery capability determined.

11. Serviceability testing. System maintainability criteria as they relate to dump programs, trace programs and diagnostics for example are part of system testing. Service support documentation must be assessed to determine man-machine compatibility. Maintenance efficiency is important to the achievement of system availability.

12. Publications testing. User documentation needs evaluation as much as the system itself. This documentation should be the source for test plan preparation.

13. Human factors testing. Although this aspect should be considered in the design phase, any problems with the man-machine interface should be noted and changes made where appropriate. Any changes are likely to be minor.

14. Procedure testing. All of the procedures to be followed by the user need testing. These procedures include system operation as well as terminal operator procedures.

System testing in effect should expose the system to all aspects and to the ultimate limits of its operation. It is best performed by those not responsible for its development. As with other types of tests, what is needed is a highly destructive test sequence that will truly test the system. The only way to be assured of achieving that end is to place the responsibility with those not having a vested interest in the end product.

The next section outlines a representative software evaluation plan that reflects the foregoing philosophical perspective in a practical way.

REPRESENTATIVE EVALUATION PLAN

This policy statement establishes the strategy to be followed in a major development project which follows a life-cycle approach. Each project phase has a team of personnel performing their professional expertise. The Project Phase Team is responsible for defining and implementing test plans that will ensure project phase acceptance.

Errors in software can be largely attributed to breakdowns in communications between the business users, systems support, and the computer analyst. The functional specification test strategy is a high level identification for testing discrepancies between the automated system design, the functional requirements, administrative procedures, and the original objectives of the system.

Communications

To plan requires communication between the users, business systems analyst, computer analyst, and data base personnel. Upon completion of the functional specification, the Project Management Office (PMO) must assign a business systems analyst to support the user, computer analyst, and data base personnel. The link between personnel must be formalized through specific responsibilities and deliverables, and controlled by the PMO.

Next Phase Review

The Automated Systems Design Team must first review the document for functional understanding and then write an introduction, which defines what the team feels are the goals of the module system. The effort includes:

1. Data entry. On-line interactive and batch data handling, transactions for tracking sequence of events, and data element and record relationships.
2. Business operations. Functionality and procedures in place to handle automated system processes.

3. Reports. Definition, content, business operation relationship, and distribution.
4. Interfaces. Manual or automated communications and data transfer to other business systems (e.g., Material Control, Field Service).
5. Financial requirements. Incorporated and functional within the business operation.
6. Conversions. Data access defined, reports, forms computer programs and operational requirements (i.e., storage location labels).
7. System as a whole. High level function flow of all business operations.

Upon completion of the above first phase, the business systems analyst (BSA), computer analyst, and data base personnel should conduct workshops to validate their findings, to answer questions, to clear up misinterpretations, and establish an understanding of the business functionality. The end result of the workshops should provide a clear understanding for translating all business requirements that are to be automated from the functional specifications into technical requirements. The next step is to refine the automated functionality into further detail for the business programming personnel.

Functional Detail

Further detail will result in a functional document containing the sections described below.

General systems flow. The general systems flow will identify general systems process concepts and flow. This section talks about processing controls (i.e., total processing functions, responsibilities, and operation controls), data control, and integrity to ensure that the data can be controlled, accounted for, and it is accurate. Major records and functionality should be listed, along with macro system flowcharts identifying the different process flows.

Policies. This section will identify the group responsibility for defining and implementing policies and procedures that will govern the processing of a business system module. Procedures should be listed and reviewed with the user community. The policies and procedures document should run concurrently with the automated design.

Control. This section will describe the controls utilized by the business system module (i.e., transactions, reports) and relate them to the general system flow and policies.

Output. This section will describe the various outputs of the system, illustrate all the reports, and obtain user approval of reports.

Input. This section will describe the various input documents used by the system (i.e., screens, forms), illustrate all screens and forms, and obtain user approval of the inputs.

Records. This section will describe all records used in the system. The description will include:

Index of records
Prefixes
Copy members

Data elements. This section will provide a review of all data elements and relationships identified in the functional specification data model to the general flow, output, input, and records. The Automated Systems Design Team will make the necessary corrections and identify the reasons to the BSA, and data base personnel.

System transactions. The Automated Systems Design Team will explain the processing functions of the various system transactions (i.e., additions, changes, deletions, inquiries, receipts) and relate these to the general system flow.

Automated design halfway point. This should be considered the halfway point in the automated design. A workshop should be called and attended by the BSA, computer analyst, data base personnel, and computer programming supervisor. The computer analyst will identify the automated design and its relationship to the functional specification business requirements and data model. The BSA will ensure that business requirements functionality is being provided. The data base personnel will review records and data item relationships to ensure that the data model being designed is workable and acceptable. The programming supervisor will review the system and begin to formulate plans for the programming phase.

The end result of this workshop will provide communication, team association, and assurance to management and the user community that the project is proceeding in the proper direction and that business requirements are being automated.

Final Phase

The final steps in the automated design will include the following functions.

Detail processing and program specification. The Automated Systems Design Team will describe the detail of the system required for programming personnel. This description will include:

Index of processes
Index of job streams/programs (e.g., process, job stream, program)
Index of files (e.g., identification, description, type, device, source, destination)
Detail process flow
Detail program specification (e.g., description, input, output, detail process, computations)

System interface. The Automated Systems Design Team will define the interface between the business system and other systems. The interface will be identified in descriptive and pictorial illustrations.

System requirements. The Automated Systems Design Team will define the resource requirements for the development and continual production support of the system. The resource requirements will include:

Hardware resources
Software resources
User/system manpower development
User/system manpower production

Security/backup. The Automated Systems Design Team will detail the security requirements of the system. The security requirements will address record backup, history, fiche, reports, forms, and program backup.

System constraints. The Automated Systems Design Team will determine any constraints that will be placed on the user sites (i.e., site interface, data center priority).

Responsibilities. The Automated Systems Design Team will describe the responsibilities of the various groups involved in the system functioning.

Test requirements and plans. The Automated Systems Design Team will describe the testing procedures to be used during the remaining phases of the testing cycle. These procedures will address unit testing, integration testing, system testing, security testing, and testing of user manuals.

Unit Testing

The first phase of testing is performed by the programmer. The testing is performed with a low volume of data and each routine in the program is tested. The results will have been predetermined. All discrepancies between predetermined and actual results must be resolved.

Integration Testing

The second phase of testing is also performed by the programmer. This phase of testing is performed with a low volume of data. The purpose is to determine that the programs "fit" together and that data is passed correctly from program to program. Again the results will have been predetermined. All discrepancies between the predetermined and actual results must be resolved.

System Test

The third and final phase of testing is done by the programmer. The system test will be subjected to reasonably large amounts of input data. This data will be entered into the system at the same time to demonstrate how the system will perform under peak loads. The purpose is to determine that the programs "fit" together with volumes of data, that volumes of data can be successfully passed from program to program, and that the sheer volume of data does not cause any defects in the system.

Acceptance Testing

Acceptance testing is performed by the user community. This comprehensive testing is done with a large volume of data and is a simulation of the production mode of operation. All manual procedures, data center procedures, distribution, and function requirements will be tested. The acceptance of the system by the user and the data center is the end result of this phase of testing.

Security

Testing will be required to ensure that the system does not allow users to perform functions higher than they are authorized to perform.

User Manuals

User manuals will be tested to uncover areas where they do not accurately reflect the system.

Walk through and communications. Upon completion of the final detail specifications, a workshop is required for a final evaluation of the automated design. At his point a senior programmer analyst should be involved to perform analysis on the program flow, detail specifications, and data base for initiation of the programmer phase. The computer analyst should walk through the detail flow with the BSA, data base and programming personnel. The end result of the workshop is a final approval that the systems automated function can continue to the next project phase, and that it is meeting the original objectives of the system.

The following section carries the foregoing strategy one step closer to the implementation, i.e., it presents the outline of a software test plan.

A TYPICAL SOFTWARE TEST PLAN

The following test plan defines the total scope of the testing to be performed on a specific project. It identifies the particular level of testing and describes its contributing role for insuring reliability and certified acceptance of the computer program. Individual test requirements are listed for every test to be conducted at the specified level of testing. The test plan contains precise statements of the purpose, scope and schedule for the individual test being planned. It identifies the degree of testing and the specific functions that are involved in the test. Also, the specific objectives of the test are defined and a summary of the test methods and the type of system environment to be used are included.

Scope

This software verification program includes the following test levels:

Unit Test

A unit is a group of logically related software routines. The intent of the test is to ensure that all branches in the routines are checked. The environment external to the unit will be simulated.

Software Subsystem Test

The system software program consists of the Mission Management Subsystem (MMS) and Signal Processor Subsystem (SPS). All units within a subsystem are integrated and tested as applicable, to verify the man-machine interface, control of peripheral devices, data entries from peripheral devices, data files stored on the disk, and program restarts. The test environment will include operational hardware as much as possible. Interfaces will be simulated with test tools for those cases where the operational equipment is not yet available.

Software System Test

After the software has been validated at a subsystem level, MMS and SPS software will be integrated and tested in a master MCS and a slave MCS configuration. As with the subsystem tests, operational hardware will be used wherever possible. Test cases executed in the final stages of software system test will be performed in a multiset environment simulating a system composed of a master MCS, slave MCSs, and RSSs.

When software testing has been successfully concluded (tests cited above), the software is approved. At the same time, the software is turned over to the System Test and Installation (T&I) Group for additional software/hardware integration and system acceptance tests. System T&I objectives are not addressed in this document since they are beyond its scope. During System T&I testing, software personnel serve in a support role to analyze test results and resolve software problems which may be encountered. A subset of the System T&I tests constitute the official software acceptance test.

Objectives

The objectives of the System Software Verification Program are:

1. Verify software performance in accordance with the Program Performance Specification (PPS) which is based on the System Technical Description (TD).
2. Promote a systematic approach to system integration and testing.
3. Validate the implementation of the design.
4. Identify and resolve design and coding errors.

Testing Requirements

This section provides a description of the objectives and interrelationships of the required software test levels, delineates test case requirements, and provides traceability from the PPS requirements to the allocated test cases.

Software Test Levels

Unit Testing

The *unit test* is intended to verify the operation of each subprogram or module. The environment external to the unit is simulated in order to verify each path within the subprograms or modules. In addition, the unit test verifies that the unit satisfies or supports the intended requirements and conforms to the program design. At a minimum, unit testing shall be performed to:

Verify all computations using both nominal and extreme values.
Verify all input options, output options, and formats, including error and information messages.

Test all options, output options and formats, including error and
information messages.
Ensure that each module fully satisfies (wholly or its assigned portion)
the performance and design requirements.
Ensure the capability to properly handle erroneous inputs.

Software Subsystem Testing

The software subsystem tests require that the software units be
operationally combined. The intent of this testing is to ensure that the
subsystem fulfills the requirements set forth in the PPS and the Program
Description Document (PDD). At a minimum, subsystem tests shall be
performed to:

Ensure error-free linkage of units
Ensure that the subsystem fully satisfies the detailed performance and
design requirements
Exercise the subsystem in terms of input/output performance with the
results satisfying the applicable performance and design
requirements
Ensure the capability of the subsystem to properly handle erroneous
inputs
The testing will be performed in a top-down manner by creating a
stub version of the subsystems. As units become available, the stubs
are replaced by the actual units. This approach enables a building-
block method for the integration of the subsystems and allows for
scheduled releases of individual units to be integrated in a timely
manner. The testing initially integrates the software units in critical
paths (i.e., operator interface, disk file management, device handlers,
etc.). As testing progresses, other units are integrated until the entire
subsystem is integrated. The unit development schedule is described
later.

Each subsystem will be integrated individually with the use of
simulation programs to simulate the other subsystem. During this
phase, any interfaces with any hardware elements which are not
available will simulated.

Any discrepancies discovered will be documented in a problem
report.

Software System Testing

The software system tests require that the software subsystems be
operationally combined. The intent of this testing is to verify the
proper operation of the software and to validate system performance

requirements. The testing repeats the subsystem tests without the subsystem simulation programs. As with the subsystem tests, the software system tests utilize all interfaces with hardware elements which are available. Any interfaces which are not available will be tested during System T&I which is not discussed here. Any discrepancies discovered will be documented. At a minimum, system tests shall be performed to:

Ensure that the system fully satisfies the detailed performance and design requirements
Exercise the input/output performance of each set type in a system configuration
Ensure the capability of the system to operate in a minimum and maximum configuration
Ensure the total man-machine interface
Ensure the capability of the system to properly handle erroneous inputs

Following the system testing phase, the software is released to the System T&I Group for further testing and preparation for acceptance tests. Any discrepancies noted by the System T&I Group are reported in problem reports. The problems are resolved by the developers and the updated software is released to the System T&I Group for retesting.

Test Case Requirements

Software subsystem and system test case requirements are presented in the Software Test Specification. These test cases form the basis for detailed test planning. A top-down methodology is utilized to ensure verification of all requirements while minimizing the overall testing required. Each requirement is analyzed to select the verification method and test level best suited for the requirement. At each level, tests are established to demonstrate multiple capabilities and thereby keep the number of tests to a minimum. Functional scenarios are also developed during this process to ensure early verification of critical interfaces, provide increased confidence that components will properly execute at the subsystem and system levels and to demonstrate system performance requirements.

Test requirements will continue to be reviewed and analyzed throughout the software verification program. If an assessment is made that an initial allocation is either incorrect, or the test did not demonstrate a requirement sufficiently, the allocations will be revised and test procedures adjusted. Formal approval of such test modifications will be provided by a software Test Review Board (TRB).

Test Traceability

The basic objective of the software verification program is to provide an error free product that performs in accordance with the PPS. To verify that all software requirements have been satisfied, traceability has been established between the PPS requirements, the software tasks that satisfy each requirement, and the associated validating test case numbers. This traceability data, the assigned test level and verification method are provided in matrix form in the Software Test Specification.

ORGANIZATIONAL RESPONSIBILITIES

The objectives of the test effort are to verify the ability of the software system to satisfy the requirements specified in the PPS and to demonstrate that the system performs satisfactorily over a specified operational scenario. To accomplish these objectives, support is required from several organizations. The responsibilities of each organization are described in the following paragraphs.

Producer Responsibilities

Software Test Organization

The software test team is managed by a Software System Engineering and Test Manager. He is responsible for all software test activities from the software test planning phase through turnover to the System T&I Group. It is the responsibility of the Software Test organization to generate the software Test Plan, the Software Test Specifications, and the Software Test Procedures. This group also performs the required subsystem and system tests and prepares the Software Test Report. Figure 1 shows a typical schedule for the required test documents:

Document	Prelim	Draft	Update	Final
Test Plan	Aug	Jan	Jun	Mar
Test Specification		Jan	Sep	Mar
Test Procedures		May	Jun Sep	Mar
Test Report				Mar

Figure 1 Test Document Schedule

System T&I Group

The System T&I Group is responsible for ensuring that the hardware configuration for the various phases of the software testing is available and operation. This organization is also responsible for assisting the software personnel in the setup and maintenance of the hardware as applicable.

Customer Responsibilities

The customer is responsible for reviewing and approving the test plan, test procedures, and test results.

PERSONNEL REQUIREMENTS

The Software Test organization consists of a Software Test Manager (Software System Engineering and Test Manager) and other personnel who participated in the design and development of the software system. The Software Test Manager is responsible for the generation of the test plans, procedures, specifications, and reports. The manager will also participate as a member of the TRB. All test team members will report to the Software Test Manager who will direct and monitor the activities for each phase of testing.

The responsibilities of the test team are to assist the Software Test Manager in defining simulation programs required for the software subsystem and system tests, identify test cases for the software subsystem and system tests, perform the software subsystem and system tests, and resolve all discrepancies reported during the testing phases.

A proposed staffing profile for this organization is provided in Figure 2.

HARDWARE REQUIREMENTS

The system consists of a computer and a signal processor (consisting of four computers). In addition, a development system (consisting of a computer, a 190 mega-byte disk, a line printer, a magnetic tape unit, and user terminals) will be utilized to develop the code. A cross-assembler or cross-compiler hosted on a VAX 11/780 will be used to develop the code for the four computers. The computer availability requirements during the testing phase is as follows:

1. Unit Test Phase. The development system with the addition of a plasma display(s) and keyboard(s) will be used to unit test the Mission Management Subsystem software units. The Signal Processor Subsystem software units will be unit tested on the VAX 11/780 using a simulator.

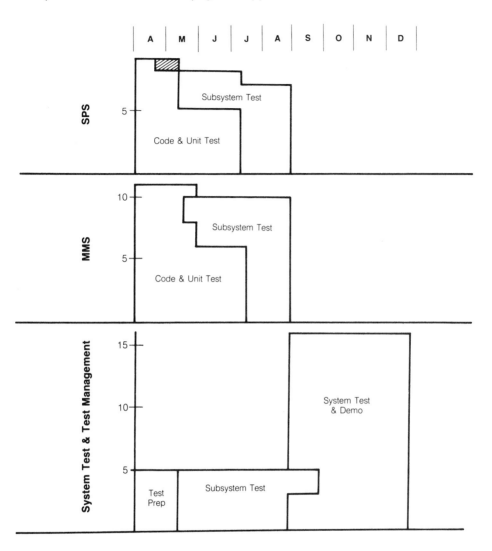

Figure 2 Test Organization Staffing Plan.

2. Subsystem Test. The Mission Management Subsystem, which executes in the computer will be tested utilizing the 35.6 mega-byte 8 inch disk, the plasma displays and keyboards. The signal data processor interface, the received control display units interface, the received control display units interface, the fault indicator interface, and the Signal Processor Subsystem interface will be simulated during MMS subsystem test. The Signal Processor Subsystem, which utilizes four computers, will be tested utilizing the actual signal processor hardware.

3. Software System Test. The Software System Test will integrate the Software Subsystem and the hardware elements which are available. In addition, a multiple set configuration will be simulated to verify the operation of the complete system.

Detailed hardware requirements for individual test cases will be incorporated in the Test Specification Document.

SUPPORTING SOFTWARE REQUIREMENTS

Program development utilities include text editors, assemblers, compilers, linkers, program libraries, and debuggers. In addition, programs to simulate the Mission Management and Signal Processor subsystems will be developed to aid during unit and subsystem testing. Simulation programs will also be required for hardware devices which are not available.

Test software also includes stubs to simulate modules which have not been integrated into the system. The stubs provide a means to integrate the software modules in a subsystem by starting with a nucleus of modules (i.e., operator interface functions, command interpreter functions, disk file functions, general utilities, etc.). As other modules complete the unit test phase, the respective stubs are replaced and the integration process continues until all of the subsystem's software is integrated.

Detailed simulation requirements and their scheduled need dates will be incorporated in the Test Specification Document.

SCHEDULE

The integration and test schedule is structured to enable the hardware and software tasks to be completed in a timely manner. As shown in Figure 3, the software testing phase begins approximately April 19 and continues until December 19.

Units will be developed with the critical path units (i.e., operator interface, file handlers, device handlers, and drivers, etc.) scheduled for the earliest unit test completion in order to build a nucleus for the following test phases. The units will be integrated incrementally as they complete the unit test phase to permit early testing of critical components, minimize coding of test drivers and allow early exposure of any interface problems. The detailed subsystem and system test schedules are contained in the Software Test Specification.

Management and Control

Management and control mechanisms will be established to support top-down implementation of the software integration and test program.

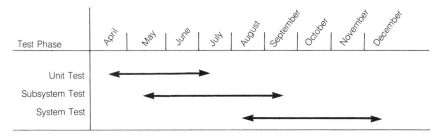

Figure 3   Software Test and Integration Schedule

This section describes the management mechanisms to be utilized during integration and test to assure adequacy of testing and effective control.

Roles and Responsibilities

Unit tests and preliminary integration testing will be performed by the responsible designer/programmers under the direction of their subsystem development manager. These tests are conducted informally but are still controlled and require that the test planning, test procedures, and test results be kept available for review. Management of the formal software subsystem and system tests will be the responsibility of the software test manager.

Test Review Board

A software TRB will be established to review the adequacy, accuracy, and completeness of unit, subsystem, and software system test phases. This review board will consist of the Software Test Manager, Software Development Manager, and Configuration Control Manager. The purpose of the TRB is to provide an independent assessment of the software test activities (starting with a review and approval of test plans/test procedures and ending with a review of detailed test case outputs/results) and act as the internal certification/acceptance agent which grants authority to proceed to the next test phase.

Software Turnover Criteria

This section presents software turnover criteria for the unit, subsystem, and system test phases.

Unit Test Turnover Criteria

The criteria for completion of unit testing include:

Completion of unit tests
Completion of the Unit Development Folders
Review and approval of the unit testing by the subsystem development
    manager
Review and approval of the unit testing by the TRB

Subsystem Test Turnover Criteria

The criteria for completion of subsystem testing include:

Completion of planned subsystem test cases
Demonstration of the unit interface capability
All requirements allocated to the subsystem test level have been verified
Completion of the test case folders for the tests performed
Completed update of the test traceability/status matrix
Review and approval of subsystem testing by the Test Review Board

System Test Turnover Criteria

The criteria for delivery and approval of the software upon completion
of system testing shall include the following:

Quality assurance signed confirmation that system test performance
    satisfies the acceptance criteria for each test case that applies to the
    delivery
Completion of test case folders for the system tests
Completed update of the test traceability/status matrix
Completion and turnover to configuration management of the Test
    Review Board review package including a summary of known
    problems (open Software Discrepancy Reports), and a schedule for
    closeout of all outstanding SDRs
Transfer of all TCFs to configuration management.

Test Review/Approval Process

When discrepancies are detected during testing a Software Discrepancy
Report is generated. SDRs will be analyzed by the TRB and categorized
as a problem, a nonproblem, or a product improvement. All
discrepancies determined to be a problem are assigned to test team
members to be resolved. When the discrepancies are resolved, the Test
Review Board evaluates the solution, and unit retest results and

determines the extent of retesting required at the subsystem or software system test levels.

Following the completion of each level of testing, the test case folders will be distributed to members of the TRB for review. Subsequently, a TRB meeting will be conducted to determine if all the requirements allocated to the test level were adequately demonstrated and whether any problems encountered were of a significant nature.

Change Control

Prior to unit turnover the control of changes to the software products resides with the software development manager. Following the completion of the unit development phase, the unit is released by the developer and is placed under formal configuration control. Changes to software products after being placed under formal control will be permitted only through the use of formal change procedures described in the Software Configuration Management Plan.

Retest Criteria

Retesting requirements for software changed as a result of Software Discrepancy Reports will be approved by the Test Review Board. If a task or module is redelivered when turning over a new software capabilities, retesting will be required at the unit level to verify that the original capabilities have not been affected. This retesting will be required in addition to the unit tests designated to test the new capabilities.

Traceability

Test evaluation traceability documentation will be utilized as a management tool to provide current test status and insure that program performance specification requirements were implemented and tested. This documentation will provide traceability from the original requirement to the applicable software modules and then to the allocated test level and test case. Data will also be maintained on the method of verification, current verification status and the status of all discrepancy reports.

Documentation

The following documentation will be used for software testing activities in addition to this Software Test Plan.

Software Test Specification

This document provides a test specification in two parts. Part 1 shall be the System Test Specification and Part 2 shall be the Functional Test

Specification. All tests that are to be used for subsystem and system tests shall be identified, including a description of input, output, operator actions and other pertinent test related information. It will be updated to reflect detailed test data values and used as a basis for the development of test procedures.

Software Test Procedures

This document provides the setup, execution, and evaluation steps for each specific test identified in the test specification.

Unit Development Folders

A unit development folder (UDF) is maintained for each software unit. The purpose of the UDF is to provide an organized, accessible collection of the requirements, design data, code, test data, and any developer notes pertaining to the unit. In addition, all problem reports and resolutions are maintained in the UDF. Each UDF is reviewed at periodic intervals by the software test manager, the software development manager, and software quality assurance to determine the progress of the unit development. Figure 4 illustrates an outline of the UDF.

Unit test cases will identify support software requirements (e.g., test data generators, test drivers, etc.) and state how the test will verify compliance of code with design, the intended functional capabilities, and unit level design budgets (sizing, accuracy, and timing).

Test Case Folders

Test case folders containing data pertinent to each test case will be compiled and maintained for all subsystem and system tests. These folders will consist of the following items:

1. The test procedure to be followed during the test case execution. If modifications are made during execution the procedure will be annotated to represent the as-run operations. The annotated procedures will be initialed at each correction by the test conductor and the test monitor.
2. Test execution report for each run made of the test case. The test execution report will be completed by the appropriate individual each time a test is attempted, whether the attempt is successful or not. It will be used to historically document all attempts of each test case and what transpires during each of those attempts.
3. Hardcopy output from the final test execution (where hardcopy capability exists).

4. Copies of any Software Discrepancy Reports generated as a result of a test execution and their final disposition.
5. Test results/analysis reports will be used to document the extent to which planned test objectives were successfully completed and any recommendations that may arise from the test analysis including recommended document revisions, additional tests and revisions to test plan/test procedures.

Software Test Report

This document provides a final report summarizing quantitative results of all formal tests. It will be produced at/shortly after formal software acceptance test completion.

SUMMARY

Software testing has been presented from both a philosophical and practical viewpoint. The various levels of testing were identified and the role each plays in software evaluation discussed. Also presented were a representative evaluation plan and a typical software test plant.

RECOMMENDED READINGS

Cooper, John D., and Matthew J. Fisher, editors. *Software Quality Management.* New York: PBI, 1978.

Glass, Robert L. *Software Reliability Guidebook.* Englewood Cliffs: Prentice Hall, 1979.

Dunn, Robert, and Richard Ullman. *Quality Assurance for Computer Software.* New York: McGraw-Hill, 1982.

NOTE

[1] Glenford J. Myers, *Software Reliability Principles and Practices.* (New York: John Wiley & Sons, 1976), pp. 231−234.

Chapter 13

Sources of Product Evaluation Data

This chapter deals with the sources and utilization of technical information available to manufacturers from technological advances resulting primarily from government contracts and subsidies. Recommendations for methods of identification, evaluation, and dissemination of new concepts that have meaningful industrial potential are outlined. Many sources of application information, their processing system, preparation, and distribution are listed. Included are sources of information pertaining to improved processes and techniques (for example, new ways of fabrication, forming a part, or scheduling a job). Described in detail are the many facets of the Defense Technical Information Center, the Government-Industry Data Exchange Program (GIDEP), and the National Referral Center.

DEFENSE TECHNICAL INFORMATION CENTER

The Defense Technical Information Center (DTIC) (formerly the Defense Documentation Center) is the central facility within the Department of Defense (DOD) for processing and supplying scientific and technical reports of Defense-sponsored or cosponsored research, development, test, and evaluation efforts in the form of planned, ongoing, and completed research. This information is put into an on-line data base-(the Defense RDT&E On-Line System (DROLS)-and made available to U.S. government organizations, government contractors, grantees, and participants in the DOD potential contractor program.

DTIC's principal products include the following:

Full-test copies of technical reports from DTIC holdings.
Automatic Document Distribution (ADD) Program

Automatic distribution (every two weeks) of microfiche copies of newly accessioned scientific and technical reports in accordance with user subject-interest profiles.

Current Awareness Bibliographies (CAB). This program matches the user's subject interest profile against newly acquired documents. The end product is a technical report bibliography sent to the user on a biweekly basis.

Subject Bibliographies. DTIC provides demand bibliographies and data base summaries on request in responge to specific area-of-interest requirements.

Recurring Reports. Compiled from the work unit, and independent research and development data bases. Current scientific, technical, and management information supplied monthly, quarterly, semiannually, or annually, according to user requirements.

They supply a variety of publications which include:

Technical Abstract Bulletin (TAB) and Indexes. A biweekly publication (classified confidential), announcing the availability of research, development, test, and evaluation documents acquired by DTIC. It is distributed to authorized DTIC users (having facility clearances) without charge. The Bulletin is indexed by the TAB Indexes which are published with the TAB.

Source Hierarchy List. A two-volume hierarchical listing of related organizations and codes used in the DTIC data bases (AD-A120 000, Vol. 1; AD-A120 001, Vol. 2).

Source Header List. A two-volume listing of source names and their codes used by DTIC in its data bases (AD-A115 000, Vol. 1; AD-A115 001, Vol. 2).

Government acronyms and alphabetic organizational designations used in DTIC. A guide to government acronyms reflecting reports processed into the DTIC collections (AD-A124 500).

How To Get It—A Guide to Defense-Related Information Resources. A reference tool identifying government information resources of interest to the defense community (AD-A110 000).

DTIC Referral Data Bank Directory. Listings of mission, scope and services for government-sponsored activities which provide scientific and technical information services (AD-A095 600).

DRIT (DTIC Retrieval and Indexing Terminology). DTIC's vocabulary for indexing and retrieval of scientific and technical literature (AD-A068 500).

DTIC Cataloging Guidelines. A cataloging manual for technical reports based on COSATI standards (AD-A080 800).

Data Element Dictionary. The dictionary standardizes the data elements and identifies the data uses that will constitute DTIC's uniform system (AD-A083 800).

Notices of changes in classification, distribution, and availability; annual cumulation. This DTIC cumulation contains change entries for all reports that are downgraded, declassified, delimited, and/or transferred to NTIS for public sale. Available to registered users. Published annually since 1976.

The collection contains over 1.2 million technical reports under computer control and an additional 300,000 documents available for manual searching.

The DTIC data base consists of bibliographic citations and related summaries concerning planned, ongoing, and completed research activities.

The four data bases are:

1. The Research and Development Program Planning Data Base (R&DPP) consists of projects which forecast and propose future research efforts. Input was discontinued as of January 1, 1983. However, the existing data base is still available, and a replacement is under consideration.

2. The Research and Technology Work Unit Information System (WUIS) contains research projects at the work unit level that are currently being performed by DOD and NASA, or under DOD contract.

3. The Technical Reports Data Base (TR) consists of descriptive summaries for reports on completed research efforts.

4. The Independent Research and Development Data Base (R&D) contains research projects currently in progress in industry, which may have future applications to—and compete for—DOD contracts. This data base contains proprietary information and is available only to Department of Defense organizations.

DTIC's collection is specialized and includes areas normally associated with defense; such as aeronautics, missile technology, space technology, navigation, and nuclear science. However, DOD's interests are widespread, and include such sectors as biology, chemistry, energy, environmental sciences, oceanography, computer sciences, sociology, and human factors engineering.

Other "special" parts of the collection include:

A referral data base, that contains references to other data collections and sources of information.

World War II documents, entitled ATIs (Air Technical Index), are also a special part of the DTIC collection.

DOD-sponsored patents and patent applications on file are another special collection element.

There are many users of this vast collection of evaluation data. Some of the major users are listed below.

Libraries comprise a substantial number of DTIC's active users. They include: university libraries, special libraries, and even the Library of Congress.

Department of Defense agencies such as the Army Corps of Engineers, Air Force Institute of Technology, and the Naval Surface Weapons Center.

Other U.S. government agencies such as the U.S. Coast Guard, the Central Intelligence Agency, the Department of Agriculture, and the Treasury Department.

Commercial companies such as Union Carbide Corp., U.S. Steel, General Electric, Boeing Corp., and IBM, as well as many smaller businesses and individuals contracting with DOD and other federal agencies.

Participants in the potential defense contractor program.

The Department of Defense offers a special program to provide technical enformation support to organizations or individuals that do not have current contracts or grants with DOD. Through the potential defense contractors program, industrial companies, educational institutions, nonprofit technical organizations, and separate individuals, with adequate research and development capabilities, are given access to the technical reports collection and data files maintained by the Defense Technical Information Center.

Registration for this program must be made through one of the military services or DARPA since each has guidelines, forms, documentation, and other informational material pertaining to individual programs.

Points of Initial Contact

For particulars concerning registration, directories of contact points, and other information for each program contact one of the following:

Air Force Potential Contractor Program (PCP)
Air Force Information for Industry Office
Tri-Service Industry Information Center
5001 Eisenhower Avenue
Alexandria, VA 22333
(202) 274-9305
Autovon 284-9305

Army Qualitative Requirements Information Program (QRI)
Commander
Army Material Development & Readiness
Command (DARCOM)
Attn: Technical Industrial Liason Office (TILO)
5001 Eisenhower Avenue
Alexandria, VA 22333
(202) 274-8948
Autovon 284-8948

Department of the Navy/Industry Cooperative R&D Program (NICRAD)
Chief of Naval Material (MAT 08T4)
Navy Department
Washington, DC 20360
(202) 692-0515
Autovon 222-0515

DARPA Potential Contractor Program (DARPA/PC)
Defense Advanced Research Projects Agency
Attn: TIO
1400 Wilson Boulevard
Arlington, VA 22209
(202) 694-5919
Autovon 224-5919

DOD Military and civilian students.
Universities throughout the United States.

Although organizations register for service from DTIC, individuals in the agencies are the actual users. These include engineers, scientists, managers, administrators of research efforts, and librarians.

The general public can also gain access to DTIC information. They release unclassified/unlimited technical reports and bibliographic information through the National Technical Information Service (NTIS). DTIC documents released to NTIS are indexed in NTIS's Government Reports Announcements & Index and are available on-line through the NTIS Bibliographic Data File. This file can be accessed through commercial data base vendors. For further information direct inquiries to Customer Service, NTIS, 5285 Port Royal Road, Springfield, VA 22161; Telephone (703) 487-4660.

As mentioned previously, DTIC is an on-line service achieved through DROLS. It was developed by DTIC to provide on-line access to its collection contained in the four separate data bases previously described.

DROLS is used for interactive retrieval, input, and ordering documents. Practically the entire collection can be searched and displayed provided a terminal station lined to DTIC's central computer system is available.

A wide variety of terminals are available for this purpose. These include:

Access may be gained by using the TYMNET commercial data communications network.
DROLS can communicate with any terminal (CRT or typewriter) which employs the standard ASCII asynchronous protocol. Terminal communications speeds are 300 or 1200 baud (30 or 120 characters per second) in even parity.
There is a charge per connect hour or proportionate share.
Subscribers to this service must have a deposit account with the National Technical Information Service.
Users will not be charged for time they input technical data into the DTIC data bases.
This service is limited to unclassified access only. Users requiring classified access will be required to use the specialized UNIVAC 100 or 200 CRT with dedicated telephone lines.

THE GOVERNMENT-INDUSTRY DATA EXCHANGE PROGRAM

The GIDEP (Government-Industry Data Exchange Program) is a cooperative activity between government and industry participants seeking to reduce or eliminate expenditures of time and money by making maximum use of existing knowledge. The program provides a means to exchange certain types of technical data essential in the research, design, development, production and operational phases of the life cycle of systems and equipment.

The program is centrally managed and funded by the government. Its participating organizations are: the United States Army, Navy, Air Force, Department of Labor, Defense Logistics Agency, General Services Administration, National Aeronautics and Space Administration, Federal Aviation Administration, Department of Energy, U.S. Postal Service, National Bureau of Standards, and the National Security Agency as well as the Canadian Department of Defense and includes hundreds of industrial organizations. A limited data exchange on electronic parts and components has also been arranged with the European EXACT program.

As a result of government emphasis on commercial off-the-shelf items, any activity which uses and/or generates the types of data GIDEP exchanges may be considered for membership. The program specifically excludes classified and proprietary information.

Participants in GIDEP are provided access to the four major data interchanges listed below. The proper utilization of the data associated with these interchanges can assist in the improvement of quality and reliability and reduce costs in the development and manufacture of complex systems and equipment.

Engineering Data Interchange
Metrology Data Interchange
Reliability-Maintainability Data Interchange
Failure Experience Data Interchange

The Engineering Data Interchange contains engineering evaluation and qualification test reports, nonstandard parts justification data, parts and materials specifications, manufacturing processes, and other related engineering data on parts, components, materials, and processes. This data interchange also includes a section of reports on specific engineering methodology and techniques, air and water pollution reports, alternate energy sources, and other subjects.

The Metrology Data Interchange contains metrology-related engineering data on test systems, calibration systems, and measurement technology and test equipment calibration procedures, and has been designated as a data repository for the National Bureau of Standards (NBS) metrology-related data. This data interchange also provides a Metrology Information Service (MIS) for its participants.

The Reliability-Maintainability Data Interchange contains failure rate/mode and replacement rate data on parts, components and materials based on field performance information and/or reliability demonstration tests of equipment, subsystems, and systems. This data interchange also contains reports on theories, methods, techniques, and procedures related to reliability and maintainability practices.

The Failure Experience Data Interchange contains objective failure information generated when significant problems are identified on parts, components, processes, fluids, materials, or safety and fire hazards. This data interchange includes the ALERT and SAFE-ALERT data, failure analysis and problem information data.

Organizations may participate without charge in any or all of the above data interchanges by agreeing to abide by pre-established requirements for participation.

Manufacturer Test Data and Reports. The GIDEP data base also includes certified test reports from manufacturers detailing test results and inspections conducted on devices of their manufacturer. Test data pertains to commercial as well as military and high reliability devices. The availability of this test data in the GIDEP provides participants the opportunity to profitably apply the data in every phase of system design, development, production, and support process.

A recent addition to GIDEP's data interchanges has the Aeronautical Depot Maintenance Industrial Technology (ADMIT) program which exchanges data on special equipment and processes utilized in the overhaul and maintenance of aeronautical systems and components. This data interchange is presently limited to participating government agencies.

SPECIAL SERVICES

Three special services are provided within GIDEP. The ALERT system, which notifies the participant of problem areas; the Urgent Data Request (UDR) system, which allows a GIDEP participant to query all other GIDEP participants on specific problems and the MIS which provides rapid response to GIDEP participants on queries related to test equipment and measurement services. The MIS system also includes an extensive research capability which is available to participants on a fee basis.

The ALERT system provides the GIDEP participant with identification and notification of actual or potential problems on parts, components, materials, manufacturing processes, test equipment, or safety conditions. The initiator of the ALERT coordinates the ALERT with the manufacturer (vendor) when applicable, then forwards it to the GIDEP Operations Center for distribution to all participants.

The UDR system permits any participant with a technical problem to rapidly query the scientific and engineering expertise of all participant organizations. A UDR form is initiated by the member and sent to the GIDEP Operations Center for distribution to all participants. Responses are provided directly to the person making the query and are also incorporated into the appropriate data interchange.

Utilizing the extensive metrology information and expertise available at the GIDEP Operations Center, the MIS system provides GIDEP participants with the capability to obtain technical information and research efforts on metrology and test-related requests. Requests which require efforts beyond the GIDEP base-line funding will be undertaken only with additional funding from the requesters. The research and expertise available encompasses the areas of research, development, test, and evaluation of measuring instruments and their application to all facets of support, maintenance, and performance of prime equipment. Outputs are intended to provide practical solutions to specific problems.

APPLICATION

With a little planning and initiative, the information available in the GIDEP can be profitably applied in every step of the system design, development, production, and support process. Design engineers will find a ready source of proven parts information to meet specific applications; the nonstandard parts data packages are of great value during design and parts selection; reliability engineers find the failure rate and mode information invaluable; and the continuous flow of safety and potential or actual failure experience information may preclude a system malfunction at any step of the way. Logisticians find the GIDEP information useful in projecting support and resupply requirements. Production engineers frequently find new and innovative techniques in these data interchanges to expedite operations or to reduce production costs. The most important aspect of all is the broad range of direct contacts in almost every technological area.

OPERATIONS

Since the inception of GIDEP, emphasis has been placed upon the rapid transmission of current information directly to potential users, and upon having the information readily available on demand. The philosophy is to have the information waiting for the user, rather than the user waiting for the information.

There are two categories of data distribution in GIDEP: full (A), or partial (B), depending upon the organization's needs. If a full participant, microfilmed data banks, indexes, and all associated documentation are maintained within the participant organization. If a partial participant, all program materials, including data indexes but excluding the microfilm data bank, are provided. Partial participants locate specific data in the indexes and request copies from the Operations Center. Partial participants may convert to full participation status when their degree of utilization justifies their maintenance of a microfilm data bank.

Each participant, depending upon the data interchanges that are involved, submits test reports, calibration procedures, failure rate/mode data, failure experience data, and related technical information to the GIDEP Operations Center. These documents are normally generated incident to ongoing tasks or contractual requirements and are not prepared solely for GIDEP. The GIDEP Operations Center reviews, processes, computer indexes, and microfilms such documents for distribution to participants.

A manufacturer may submit certifiable test reports detailing test results and inspections on devices and equipment of their manufacturer as part of the GIDEP data base. The testing can be conducted at either the manufacturer's facilities, government facilities, or independent facilities. The test data furnished by manufacturers will afford an opportunity to provide product performance capabilities to a broader base of prospective "users" in government and industry.

GIDEP has a rapid data retrieval system which makes the microfilmed information in the data banks immediately accessible to all participants through either hard-copy indexes or the use of remote computer terminal index search. Participants can use the hard-copy indexes to retrieve specific data from the microfilm cartridges, utilizing a microfilm reader-printer. Hard copy indexes are prepared in various formats depending upon anticipated usage.

Participants having remote terminal equipment, compatible with the Operation Center's computer, may be authorized direct query access to the GIDEP data banks using a simplified operator's manual.

Data search service and other assistance in use of the program is always available by contacting the GIDEP Operations Center. A *Policies and Procedures Manual* and *Representative's Handbook* which prescribe the rules and guidelines for data submittal and in-house utilization are provided to all participants at time of entry into the program.

Participation requirements or additional information about GIDEP may be obtained by contacting the Director, GIDEP Operations Center, Corona, California 91720, Telephone: (714) 736-4677, (Autovon): 933-4677.

THE NATIONAL REFERRAL CENTER

The National Referral Center in the Library of Congress is a free referral service which directs those who have questions concerning any subject to organizations that can provide the answer.

The referral service uses a subject-indexed, computerized file of 13,000 organizations, called information resources by the center. A description of each resource includes its special fields of interest and the types of information service it is willing to provide. The National Referral Center file, which is maintained by professional analysts, is used primarily by the center's referral specialists. It also is accessible to readers at the Library of Congress through computer terminals located in various reading rooms and to many federal agencies nationwide through the RECON computer network operated by the Department of Energy; future access through other federal networks is under development.

The referral center is not equipped to furnish answers to specific questions or to provide bibliographic assistance. Instead, its purpose is to direct those who have questions to resources that have the information and are willing to share it with others. Some of these resources exist within the Library itself.

SCOPE

When it was established in 1962, the center made referrals in the areas of science and technology. Today it handles referrals in virtually all subject areas, including the arts and humanities. The center maintains systematic coverage only for resources in the United States, although its file also includes some international and foreign resources.

The concept of information resource is broadly defined to include any organization, institution, group, or individual with specialized information in a particular field and a willingness to share it with others. This includes not only traditional sources of information such as technical libraries, information and documentation centers, and abstracting and indexing services, but also such sources as professional societies, university research bureaus and institutes, federal and state agencies, industrial laboratories, museums, testing stations, hobby groups, and grassroots citizens organizations. The criterion for registering an organization is not its size but its ability and willingness to provide information to others on a reasonable basis.

SERVICES

Requests for referral services may be made by letter, telephone, or in person. Telephone requests are encouraged because they allow discussion and refinement of complex questions. The center will accept requests on any topic. When a subject is not covered in the data file, the center will attempt to locate new information resources from its extensive contacts.

In response to a request, the center will provide the names, addresses, telephone numbers, and brief descriptions of appropriate information resources. The reply usually is made in the form of a computer printout. In each case, the response is individually tailored to the specific request.

In order for the center to provide the most effective service, all request should include the following information:

Information Required. The request should describe as clearly and precisely as possible the specific information that is required. Service is facilitated if each request is limited to a single topic.

Information Resources Already Contacted. The request should indicate the information resources that are known or that have already been contacted.

Special Qualifications. The request should include any pertinent qualifications that may entitle the requester to use resources that are not normally available. These might include participation in a government contract, affiliation with a recognized project, membership in a professional society, and others.

The following is a sample of questions which the center receives:

Who can tell me about on-board production of methane for vehicle propulsion?

What organizations are currently doing research on international terrorism?

Where can I locate information on the flash point of textiles and other and other consumer products?

PUBLICATIONS

The National Referral Center occasionally compiles directories of information resources covering a broad area. These are published by

the Library of Congress under the general title *A Directory of Information Resources in the United States* with varying subtitles, and are sold through the Superintendent of Documents, U.S. Government Printing Office, Washington, DC 20402
A recent volume in the series is:

Geosciences and Oceanography 1981. 375 pages $7.50 (GPO Stock No. 03-000-0131-1)

Also sold by the Superintendent of Documents is a similar guide prepared by the center:

Directory of Federally Supported Information Analysis Center. Fourth ed. 1979. 87 pages $4.00 (GPO Stock No. 030-000-00115-0)

Under the title *Who Knows?*, the center issues informal lists of resources that have information on specific topics, such as hazardous materials, population, or environmental education. The lists are available free of charge from the center. They must be requested individually by topic (a list of titles is available).

REGISTRATION

The National Referral Center invites organizations that have information in specialized fields to participate as information resources. They may register by letter or on a prepared form, available on request from the center. All applications should include the following information which is necessary for referrals:

Subject Coverage and Specialization. An organization should describe its particular subject area in depth so that specific requests may be matched with appropriate resources.
Information Functions and Services. An organization should describe the kinds of services it is willing to provide. These may include consultation, advisory, identification, expert-witness, and testing services; literature searching; preparation of bibliographies; and/or other types of assistance.
Service Conditions and Restrictions. An organization should indicate any restrictions that apply to its services, such as fees or the need for security clearance or professional qualifications.
Automation. Because a growing number of inquiries are about computer-based collections and services, an organization should describe any data bases that it maintains or to which it has access.

Correspondence should be addressed to:

Library of Congress
National Referral Center
Washington, DC 20540

NATIONAL AERONAUTICS AND SPACE ADMINISTRATION (NASA) TECHNICAL UTILIZATION

To help PT&E scientists and engineers accomplish their missions, NASA continually collects a very large number of technical reports and journal articles. These documents relate to all phases of science and technology of PT&E involved in space exploration and aeronautics, both in the United States and in other countries. The information was gathered, in part, by NASA's own research and development activities, and by its contractors and grant-holders. A larger portion is closely related information generated by other U.S. government agencies, by foreign research organizations, and by universities and industries at home and abroad.

NASA's collection of scientific and technical aerospace documents already numbers several hundred thousand titles. Growing at the rate of more than 90,000 documents and 30,000 book titles a year, this collection represents a storehouse of knowledge of unique value, not only to PT&E technology but to other scientific areas as well.

Most NASA scientists and engineers, and those working for NASA on contracts or grants, know the scope and diversity of the information assembled and how to gain access to it. Many technically oriented individuals outside the immediate NASA community may not at present share this awareness. Yet they might derive great benefit from it. Already a considerable amount of the new knowledge, born of the needs of space exploration, has proved significant in nonaerospace applications. The purpose, therefore, is to acquaint interested individuals, who lack prior knowledge of it, with NASA's scientific and technical information system, and to tell them how to use it.

The most important single aid to the person seeking to find out what kinds of technical information NASA has accumulated, and how this knowledge may be tapped, is a semimonthly NASA publication, *Scientific and Technical Aerospace Reports,* commonly called STAR.

STAR is the basic, most widely available guide to NASA's storehouse of new knowledge. It is a comprehensive journal of abstracts and indexes covering worldwide reported literature on the science and technology of space and aeronautics. Twice a month, STAR announces the latest additions to that literature. Tight processing schedules permit all items to be abstracted and indexed in STAR shortly after NASA receives them.

Not only is STAR as current as the speed of modern printing permits, but it is notably more helpful and readable than most abstract journals. Its informative abstracts give the essence of the original reports, and its several indexes quickly lead the searcher to his specific area of interest.

To get acquainted with STAR, let us scan its principal elements.

First, for the reader's convenience, the abstracts in each issue are arranged in 75 subject categories. The scope of each category is explained in considerable detail in every issue's table of contents.

STAR's selection of subject categories, arrived at after extensive analysis of report contents and in consultation with scientists and librarians, does not purport to be formal classification scheme. It is an effort to bring together subjects of many disciplinary areas into logical groups. Like every other element of STAR, it is designed to guide the user as swiftly as possible to the general subject matter that he is interested in and to prevent his wasting time searching through irrelevant subject matter.

Each subject category in STAR has a two-digit identification number permanently assigned to it. For example,

03 Air Transportation and Safety
39 Structural Mechanics
75 Plasma Physics

These identification numbers help to simplify the publication's indexes, and are especially useful in computer processing. A characteristic subject category in STAR's Table of Contents, with notes of its scope, is shown below.

03 AIR TRANSPORTATION AND SAFETY 2293

Includes passenger and cargo air transport operations; and aircraft accidents.

For related information see also *16 Space Transportation* and *85 Urban Technology and Transportation*

TYPICAL SUBJECT CATEGORY

The figure "2293" at the end of the listing is the page number in that issue of STAR where pertinent abstracts begin. Abstracts in STAR are always arranged under their respective subject categories in the order of their NASA accession numbers. Following is a typical STAR abstract, with all its elements identified.

TYPICAL STAR ABSTRACT

The accession number is a unique identification number assigned permanently to a document at the time it is accepted into the NASA information system. Its elements need some explanation. The symbol "N" indicates that the report is unclassified and has no restrictions on its distribution. The "84" following the "N" is the year in which the report was announced in STAR. The asterisk at the end of the accession number shows that the report originated within NASA or, as in this case, was produced by a contractor working for NASA. The symbol "#" indicates that the text of the report is obtainable on sheet microfilm, called microfiche, which is described later in this chapter.

The phrase "corporate source" covers all types of report originators. It might be a NASA field center, a university laboratory, a foreign research organization, or a U.S. industrial concern.

In the citation, second line above the abstract, the abbreviation "NTIS" stands for National Technical Information Service, a U.S. government sales agency (its address is given below) where the report may be purchased. If, instead, the letters "GPO" had designated the availability of the document abstracted, the reader would know that he could buy a copy from the U.S. Government Printing Office (address is also given below).

Where price codes are given, the letters "HC" represent "hard copy," which means either an original printed copy of the report or a facsimile of the original. "MF" stands for "microfiche," the technical name for sheet microfilm ("fiche" is the French word for "card"). The price codes listed for each form of copy in the cited instance are NTIS price codes. A microfiche copy of a technical report is always less expensive than a hard copy. Microfiche offers additional advantages that will be explained later in this chapter.

Every semimonthly issue of STAR contains five indexes: Subject, Personal Author, Corporate Source, Contract Number, and Report Number. In addition, cumulative indexes are published annually. Cumulative indexes contain the same five indexes as the semimonthly issues, but with some minor changes. Each entry in the annual index contains the issue number and the page number for the citation/abstract; and the author index contains the title and primary report number. The annual subject index contains a cross reference structure that is not utilized in the semimonthly issues.

STAR's principle of indexing always leads the readers from the general to the specific. To speed them in their search, it uses direct indexing. For instance, if the user of a STAR annual index is an electronics engineer who wants to find out what new research reports deal with traveling wave amplifiers, he does not bother looking up "amplifier" in STAR's subject index but turns straight to "traveling wave amplifier." However, if he *should* happen to investigate amplifier first, he would find beneath it, among many other more specific subject terms, NT traveling wave amplifier. NT means "narrower term."

STAR uses a total of 17,000 separate subject terms for indexing. Each report abstracted is then indexed by several appropriate terms, representing the major subject covered in the report. Thus the user of STAR is advised to turn at once to the most specific index term he can think of to describe the area of his interest. That way, he will probably have the shortest hunt. However, if he happens to find nothing listed there, he should look under somewhat more general terms.

Again, to save time for the scientists and engineers trying to keep abreast of new developments in PT&E, STAR describes reports in its subject index by notations of their content rather than by listing their titles. Report titles can sometimes be more mystifying than enlightening. By using notations of content instead, STAR's subject index gives the technical man as much capsuled information about each report as possible, enabling him speedily to decide whether to look up the abstract or bypass it. (In all other STAR indexes, however, report tiles are used.)

To illustrate STAR's indexing system, let us follow a specific item— the abstract of N84-10174—through all its indexes.

This item is indexed first under three subject terms in the issue in which the report was originally announced. (It appears under the same subject terms in subsequent cumulative indexes.) Those subject terms in this case are Space Shuttle Payloads, Space Shuttles, and Spacecraft Contamination. The listing under Space Shuttle Payloads is shown below.

SPACE SHUTTLE PAYLOADS

A definition of STS accommodations for attached payloads

[NASA-CR-172223]   N84-10114

Molecular contamination math model support

[NASA-CR-170899]   N84-10174

SUBJECT INDEX

The key to the abstract's location is its sequential accession number; the accessions in STAR are in accession number order throughout each issue. (If this subject index under discussion happened to be in the annual cumulative indexes of STAR, the key to the abstract's location would be the number of the page in the issue of STAR where the abstract appeared.)

The abstract of N84-10174 is indexed next in the corporate source index, as shown below.

The point of having a corporate source index is that scientists and engineers engaged in PT&E are likely to know which firms are currently engaged in research and development work in their own field. By turning at once to the corporate source index, therefore, they may perhaps be guided more quickly to reports on new developments of special interest to them than if they searched through the subject index. For reports on work done for the government on contract, the contracting (or monitoring) agency—NASA in many cases—is not listed in the corporate source index. To do so, for NASA's case, would result in such a mass of subordinate index entries that the reader would be confused and slowed down in his search.

Martin Marietta Corp., Denver, Colo.
Molecular contamination math model support

[NASA-CR-170899]   N84-19174

Corporate Source Report Accession Title Number
Corporate Source Index

Next comes the personal author index, with an example appearing in it as shown below. Author's names are given in full with each abstract if they so appear on the original report, but only their initials are used in STAR's indexes. Since many reports have more than one author, STAR indexes up to five authors for a single report. As a space-saving measure, only the author and the accession number are contained in the author index.

Weiss, M.S.	N84-10734
Weisshaar, D.E.	N84-10247
	N84-10248
Weissman, C.B.	N84-10541
Weissman, S.H.	N84-10588
Welch, E.C.	N84-10992
Wells, J.H.	N84-10760
Wendland, W.M.	N84-10695
Wendt, R.L.	N84-10396
Wenguau, W.	N84-10512

The main purpose of this index is to enable individuals within a specific discipline to find out quickly what leading researchers in their field have accomplished recently.

After the personal author index comes the contract number index in which, as shown below, each contract number is matched with its accession number.

NAS8-34381	N84-10582
	N84-10583
	N84-10584
NAS8-34651	N84-10182
NAS8-34677	N84-10175
	N84-10176
NAS8-34945	N84-10174
NAS8-35017	N84-10181
NAS8-35339	N84-10177
NAS9-13247	N84-10166
NAS9-15800	N84-10647

After the contract index comes the report number index in which, as shown below, each report number is followed by its NASA accession number. This index is designed for the PT&E researcher who knows the number of a report that he ought to investigate but who does not have a clue as to where to find the report abstracted in STAR. The index is also useful to librarians in identifying or ordering reports. The asterisk and pound sign following the accession number mean that the document was NASA founded (*) and that the document is also available on microfiche (#).

NASA-CR-168265	N84-10662*#
NASA-CR-170591	N84-10645*#
NASA-CR-170895	N84-10182*#
NASA-CR-170901	N84-10181*#
NASA-CR-17092	N84-10171*#
NASA-CR-170907	N84-11077*#

With its informative abstracts and multiple indexes, both current and cumulative, STAR serves three valuable purposes.

1. *It is a current-awareness PT&E tool.* This semimonthly announcement medium covering the world's aerospace report

literature enables every user to keep abreast of new developments in his special area of interest.

2. *It is a current PT&E searching tool.* If the user wants to find out what has just been reported on a particular subject, or what other organizations are doing, or what a famous researcher has been doing since he switched from Firm A to Firm B, or what else is going on under a particular contract, STAR gives him a ready answer, right on his own desk.

3. *It is a retrospective PT&E searching tool.* STAR's cumulative indexes, prepared and distributed within two to four weeks of the close of every quarter, give the user a handy means of investigating, in depth, any aerospace report literature that interests him. Without leaving his office, he can conduct searches extending over several years. STAR's cumulative indexes have been issued, since 1963. (A similar journal in 1962 was called *Technical Publications Announcements.*)

Annual subscription rates for the semimonthly issues are $90 for domestic subscribers, $112 for foreign subscribers. Individual copies for semimonthly issues cost domestic purchasers $4.75 each. Foreign mailing is $5.95 for a single copy.

Annual subscription rates for the cumulative index issues are based upon pagination.

Domestic subscription rates apply to all countries in the Western Hemisphere, except as noted below.

Foreign subscription rates apply to Argentina, Brazil, British and French Guiana, Surinam, British Honduras, and all countries in the Eastern Hemisphere.

STAR and STAR Annual Indexes are available without charge to:

1. NASA offices, centers, contractors, subcontractors, grantees, and consultants, in support of their work.
2. Other U.S. government agencies and their contractors.
3. Libraries that maintain collections of NASA documents in work related to the aerospace program.
4. Other organizations having a need for NASA documents in work related to the aerospace program.
5. Foreign organizations that exchange publications with NASA, or that maintain collections of NASA documents for public use.

The reader may buy copies of NASA documents and noncopyrighted foreign documents listed in STAR from one of the following sales agencies, identified in STAR abstracts by the initials shown in parentheses:

National Technical Information Service (NTIS), 5285 Port Royal Road, Springfield, VA, 22161

Superintendent of Documents, U.S. Government Printing Office (GPO), Washington, D.C., 20402.

Documents available from NTIS are sold in both hardcopy and sheet microfilm (microfiche) forms. If original printed copies (identified in STAR by initials HC) are not available, facsimiles are provided.

Non-NASA documents announced in STAR may be obtained by the general public from NTIS or from the original source of each document listed in the STAR citation. They should be sure, however, to identify each document by its original report number, title, and author, and the NASA accession number. Commercial publications announced in STAR can usually be consulted in libraries or purchased from publishers or booksellers.

For STAR itself, the reader may gain access to most STAR-announced NASA-funded publications in the 11 Federal Regional Technical Report Centers and in the public libraries listed below.

Federal Regional Technical Report Centers

California: University of California Library, Berkeley
Colorado: University of Colorado Libraries, Boulder
District of Columbia: Library of Congress
Georgia: Georgia Institute of Technology, Atlanta
Illinois: The John Crerar Library, Chicago
Massachusetts: Massachusetts Institute of Technology, Cambridge
Missouri: Linda Hill Library, Kansas City
New York: Columbia University, New York
Oklahoma: University of Oklahoma, Bizzell Library
Pennsylvania: Carnegie Library of Pittsburgh
Washington: University of Washington Library, Seattle

Public Libraries

California: Los Angeles, San Diego
Colorado: Denver
Connecticut: Hartford

Maryland: Enock Pratt Free Library, Baltimore
Massachusetts: Boston
Michigan: Detroit
Minnesota: Minneapolis
New Jersey: Trenton
New York: New York, Brooklyn, Buffalo, Rochester
Ohio: Akron, Cleveland, Cincinnati, Dayton, Toledo
Texas: Dallas, Fort Worth
Washington: Seattle
Wisconsin: Milwaukee

NASA FORMAL PUBLICATION SERIES

NASA has six types of publications in its formal report series. Each of the six categories is based on best fitting the content of an author's manuscript to the needs of an identifiable readership.

The decision as to which of the six categories to use for a manuscript is made by the technical review committee acting with advice from the Center publications office. Final approval for publication is given by the Headquarters Scientific and Technical Information Branch (STIB).

Whichever category is chosen, the publication is assigned a sequential number by STIB, and the number is prefixed by a two-letter abbreviation representing its series.

The categories are:

Special Publication	SP
Conference Publication	CP
Reference Publication	RP
Technical Paper	TP
Technical Memorandum	TM
Contractor Report	CR

SPECIAL PUBLICATIONS

Special publications record scientific and technical information from NASA programs, projects, and missions for presentation to readers of diverse technical backgrounds. NASA special publications often are concerned with subjects that also have substantial external interest. This series includes:

1. Scientific summaries of mission results
2. Scientific photographic atlases
3. Histories and chronologies

4. Comprehensive program descriptions and retrospective assessments
5. Continuing bibliographies

Because the cost involved in producing special publications are generally greater and the distribution of these publications wider than for other publication categories, NASA headquarters exercises a correspondingly greater degree of control over them. Normally, titles selected for publication in this series are approved by STIB with the advice of the appropriate program offices or center publications officers.

CONFERENCE PUBLICATIONS

Conference publications normally are compilations of scientific and technical papers or transcripts from conferences, symposia, workshops, special lecture series, seminars, and other professional meetings. Some conference publications are preprinted to be distributed to conference participants; still others may be "landmark" CPs and as such may warrant extensive editorial treatment, custom handling through production and arrangement for wide availability through the U.S. Government Printing Office (GPO). For these, the originating NASA center should contact STIB for early planning.

REFERENCE PUBLICATIONS

Reference publications are compilations of scientific and technical data and information deemed to be of continuing reference value. This series includes:

1. Technical handbooks and manuals
2. Critical tables
3. Monographs, including those on design criteria
4. Scientific and technical textbooks
5. State-of-the-art summaries, including critical reviews of a body of scientific or technical literature
6. Technical reports that provide complete and comprehensive treatment of significant contributions to scientific and technical knowledge

TECHNICAL PAPERS

Technical papers record the significant findings resulting from NASA scientific and technical programs. Technical papers are the NASA counterpart to peer-reviewed journal articles and are subject to professional review controlled by the originating headquarters or center office. For documentation purposes, technical papers are preferred to professional journal articles because technical papers have less stringent limitations on manuscript length and extent of graphic presentation.

TECHNICAL MEMORANDUMS

Technical memorandums record scientific and technical findings that do not warrant or cannot be given broad dissemination because of the preliminary nature of the material, limited interest, or security considerations. Technical memorandums are either formally printed and given minimal category distribution, or reproduced in a limited number (about 250 copies) and distributed by the originating office. This series includes:

1. Preliminary data reports giving interim information of on-going research
2. Working papers prepared for the information of peers beyond the basic work group
3. Individual seminar or symposium presentations—individual papers preprinted for distribution at a symposium
4. Theses and dissertations written by NASA employees that relate to their work and that NASA elects to publish
5. Bibliographies of scientific or technical literature, with or without evaluation, generally in defined subject areas
6. Sponsored reports by NASA authors of work sponsored by other agencies
7. Security-classified reports, scientific or technical reports or papers containing classified information
8. Translations-English-language translations of foreign-language scientific and technical material pertinent to NASA work.

CONTRACTOR REPORTS

Contractor reports record scientific and technical findings by a contractor's or grantee's NASA-sponsored research and development and related efforts that NASA considers worthy of publication.

Low-numbered subseries contractor reports are those reporting the findings of significant work conducted under NASA contracts or grants. Considered analogous to technical papers, these are produced and disseminated in the manner of technical papers.

High-numbered subseries contractor reports are those publications that, although presenting new technical information, do not warrant broad dissemination. These are produced in a limited number for the use of the sponsoring office.

MERITS OF MICROFICHE (SHEET MICROFILM)

At the same time that NASA is processing incoming reports and NASA publications for announcement, abstracting, and indexing in STAR, it is making microfiche copies of each document.

Microfiche consists of flat microfilm negatives, approximately 4 × 6 inches in size, each holding as many as 98 page images.

Extraordinarily compact, microfiche represents a 24-to-1 reduction in size from the original report. Copies of 1000 average-length reports can fit into a container no bigger than a shoe box. This remarkable compactness makes it practical to maintain large collections of documents *locally*, where they are immediately available to meet requests. In addition to its low cost, microfiche is tailor made for compact storage and fast retrieval.

Microfiche is a major improvement over the old, relatively awkward reel microfilm, with its big, clumsy viewers, threading problems, and often cumbersome difficulty of locating desired material. Viewers usually are small and compact, and sit neatly on table tops. Microfiche cards are magnified for easy reading by simply sliding them into a position generally at the base of the viewer. Many of the libraries listed above offer microfiche viewing to the public.

Other advantages are that photocopies of individual pages or of all the pages in a microfiched report can be made quickly and easily on light-sensitive paper. Viewer-printers do this. In addition, the microfiche card can serve as a reproducible master from which duplicates of the microfiche can be made.

The public may purchase microfiche copies of nearly all documents listed in STAR, and many announced in IAA. Copies of STAR-listed documents can be bought from NTIS. Price codes are listed in STAR as part of the announcement of each document.

PUBLICATION AND SERVICES OF SPECIAL INTEREST TO NONAEROSPACE FIELDS NASA TECH BRIEFS

NASA issues a formal publication reporting technical innovations of potential value to the nonaerospace technical community. Published quarterly and containing about 125 items each. *NASA Tech Briefs* is free to engineers in U.S. industry and to other domestic technology transfer agents. It is both a current-awareness medium and problem-solving tool. Potential products, industrial processes, basic and applied research shop and lab techniques, computer software, new sources of technical data, and concepts can be found here. Though some new technology announcements are complete in themselves, most are backed by Technical Support Packages (TSPs) available on request. Further information on some innovations may be obtained for a nominal fee from other sources, as indicated. Each announcement indicates patent status and availability of patent licenses if applicable.

INDUSTRIAL AND STATE APPLICATIONS CENTERS

Another avenue by which the public, in particular, industrial firms not engaged in aerospace activities may gain access to NASA's vast storehouse of scientific and technological information is by utilizing the services of *Industrial and State Applications Centers.*

Under its Technology Utilization Program, NASA operates a network of dissemination centers which provide information retrieval services and technical assistance to industrial and government clients. The network consists of seven Industrial Applications Centers (IAC) and two State Technology Applications Centers (STAC) affiliated with universities across the country, each serving a geographical area. The centers are backed by off-site representatives in many major cities and by technology coordinators at NASA field centers; the latter seek to match NASA expertise and ongoing research and engineering with client problems and interests.

The network's principal resource is a vast storehouse of accumulated technical knowledge, computerized for ready retrieval. Through the applications centers, clients have access to some ten million documents, one of the world's largest repositories of technical data. Almost two million of these documents are contained in the NASA data bank, which includes reports covering every field of aerospace-related activity plus the continually updated contents of 15,000 scientific and technical journals.

Intended to prevent wasteful duplication of research already accomplished, the IACs endeavor to broaden and expedite technology transfer by helping industry to find and apply information pertinent to a company's projects or problems. By taking advantage of IAC services, businesses can save time and money and the nation benefits through increased industrial efficiency and productivity.

Staffed by scientists, engineers and computer retrieval specialists, the IACs provide three basic types of services. To an industrial firm contemplating a new research and development program or seeking to solve a problem, they offer "retrospective searches"; they probe appropriate databanks for relevant literature and provide abstracts or full test reports on subjects applicable to the company's needs. IACs also provide "current awareness" services, tailored periodic reports designed to keep a company's executives or engineers abreast of the latest developments in their fields with a minimal investment of time. Additionally, IAC applications engineers offer highly skilled assistance in applying the information retrieved to the company's best advantage. The IACs charge a nominal fee for their services.

The State Technology Applications Centers supplement the IAC system. They facilitate technology transfer to state and local governments, as well as to private industry, by working with existing state mechanisms for providing technical assistance. The STACs perform services similar to those of the IACs, but where the IAC operates on a regional basis, the STAC works within an individual state. In effect, the STAC program focuses on areas not normally served by the IACs, especially in the less industrialized states and among small businesses.

For further information on *NASA Tech Briefs*, Industrial and State Application Centers, and other publications and services of NASA's Technology Utilization Program, write to: Manager, Technology Utilization Office, NASA Scientific and Technical Information Facility, P.O. Box 8757, BWI Airport, MD 21240.

By special arrangement between NASA and the American Institute of Aeronautics and Astronautics, the AIAA publication *International Aerospace Abstracts* (IAA) is issued in coordination with STAR. Since IAA provides worldwide coverage of scientific and trade journals, books, and meeting papers in the field of aerospace science and technology, it complements STAR's worldwide coverage of report literature in that field. IAA uses the same subject categories as STAR and publisheds the same types of index. The two announcements journals are issued in alternate weeks, STAR on the 8th and 23rd of each month, IAA on the 1st and 15th.

Annual subscriptions for the semimonthly issues and cumulative index issues of IAA may be purchased postpaid from the Technical Information Service, American Institute of Aeronautics and Astronautics, Inc., 750 Third Avenue, New York, New York 10017. (At the same location, AIAA affords the reader an extensive reference collection of items it announces in IAA.)

Annual subscription rates for the semimonthly issues are $525 for domestic purchasers, $750 for foreign purchasers. The subscriptions for the cumulative index issues are $750, domestic and $1050, foreign.

AIAA does not sell any documents it abstracts except those that originated within its own organization. It does, however, provide photocopy services at the address given above. It also sells microfiche of noncopyrighted documents and, with permission, of certain copyrighted material, signifying the availability of microfiche by the symbol "#" in IAA after the accession number. Paper copies sell for $8.50; microfiche, $4.00.

SUMMARY

Several major sources of product evaluation and other technical information available to the reader have been given. Its use will be most useful and rewarding when utilized as the needs arise. We have shown the methods of identification, evaluation, and dissemination of new concepts resulting from recent technical advances. Described in detail were: the Defense Technical Information Center (DTIC), the Government-Industry Data Exchange Program (GIDEP), and the National Referral Center.

Chapter 14

A Forward Look at Product Evaluation

We have discussed the various aspects of the test and evaluation function with the greatest emphasis placed on the technical aspects. But there are other factors that require careful consideration. Little was said about man himself. In implementing the test and evaluation function there are personnel and human relations aspects such as selection, placement, promotion, transfer, firing, and the like. There are union—management relations, customer relations, and supervisor-employee issues. The fact that these aspects were not discussed here does not deny their existence. They are a very real part of PT&E.

SIGNIFICANCE OF CHANGES

Changes will occur in almost every facet of the product test and evaluation function for a variety of reasons. Principal changes will occur as a result of technology, computerization, standardization, reliability, safety, product interfaces, consumer requirements, and government regulations.

Computerized products and equipment are requiring new directions in test and evaluation. Automated machinery, including robotics, that boosts production to dramatic new heights is forcing a fundamental change in traditional concepts of product testing. Emphasis is shifting from the testing of finished products toward greater in process testing of materials components and subassemblies.

The necessity for this new direction in product evaluation procedures becomes apparent and is obvious to those involved in engineering and mass production operations. Computer-aided design and manufacturer (CAD/CAM) have the potential of turning out vast amounts of poorly designed costly scrap before anyone is aware of anything being wrong. Computer-aided product evaluation (CAPE) offers effective safeguards against their occurrence.

One obvious solution is to perform increased testing of materials being fed into the process. Proper control at this point may eliminate 50% of CAM-related problems. For instance, in paper making, such properties as basic weight and moisture content are continuously measured during production. In the future, other basic properties of paper such as opacity and brightness will be measured during in-line production. Other areas include the computerized process control of cooking and bleaching of pulp and the automatic detection of flaws during manufacturing and finishing.

Another solution is the systems approach, that is, to include in the design of the manufacturing equipment provisions for the testing of the product as an integral part of the total process. Again this was demonstrated in the examples cited above and elsewhere in the book. Management interest in this area will become keener as a direct function of the increased cost of scrap losses.

The computer has already found significant application in the product evaluation field. Its application extends from the performance of the test to the analysis of results. One chief reason for the current applications and impetus for future CAPE expansion are the economics realized. This factor insures future effort in the development of additional applications.

Current applications of the computer do not go so far as to permit "computer judgments" of test results. However, the computer does aid judgment by recording more information, identifying smaller differences, and reporting with less data more consistent results. It also has the human-factor advantage of not getting bored; nor does it invalidate the test through negligence. Consequently, it consistently keeps better records.

Perhaps most important (cost savings notwithstanding), computerized tests produce information faster, permitting more timely and informative decision making. Also computers reduce the minimum time required and reduce computational and plotting time significantly. Computers are used extensively in the process, aerospace, and automobile industries. The increased availability of time-shared computers and microcomputers permits anyone who has set up a test to have access to a service hookup or to computerize the setup on location.

EXTERNAL INFLUENCES

Government

The National Bureau of Standards in Washington has its Office of Engineering Standards Liaison engaged with other standardizing groups in developing performances specifications and test methods for industrial products. This is in response to the Department of Commerce Panel on Engineering and Commodity Standards recommendation for greater federal participation in developing and controlling the number of standards. As time goes on, industry will continue to feel the influence of government agencies on the product test and evaluation function.

Reliability

Man's concern for reliability is age-old. The key to failure-free operation is sound design, verified and controlled through extensive design reviews, failure mode and effects analysis and testing. The first automobile was quite simple but was notoriously unreliable. The automobile of today is far more complex but carries a warranty. This has been the result of extensive product evaluation and improvement activity.

Computers, aircraft, and aircraft engines are familiar products in which testing is a key factor. Less well known is the testing behind the warranty on typewriters. Several government contractors have an extensive PT&E program on their consumer lines that parallels their military products.

In the future, product reliability will be a labeled characteristic on large numbers of products, thereby permitting the consumer to select the appropriate life characteristics that he wishes to pay for. This is already possible with the ordinary electric light bulb. Most food products now carry "last sale dates" on them. All of this means more PT&E in terms of time and quantity.

Safety

Concern for product safety is a growing and lasting one. This concern is adding a new dimension to testing. Tests must be run to evaluate product performance when exposed to abnormal conditions. It also is requiring product evaluation under conditions simulating more accurately the actual use environment. For instance, most tests of tire cords show only how the tire will perform when the car is standing in the garage. Tire manufacturers now test tires in a way that simulates high-speed driving conditions. The public concern for safety will reflect itself in all products where there is a human-machine interface.

Related Fields

The influence of techniques, concepts, and equipment in other fields will also influence product evaluation efforts. New product applications will require entirely new tests. The addition of new marketing areas will require the establishment of different test and performance criteria. For instance, in the paper industry, higher speed envelope machines have made it necessary to make envelope paper with greater emphasis on the properties of stiffness, finish, and the absence of curl. The dampening systems of many lithographic presses now use alcohol and water rather than water alone, with the result that lithographic plates and blankets operate with less moisture. This means that greater attention must be given to the manufacture of paper, free from all surface debris and with coatings that will not deposit on the blanket. Computers that print information at increasingly rapid speeds on paper require tighter specifications on smoothness, stiffness, body, and the ability to refold.

Optical scanning systems require paper made to minimum brightness and stiffness levels and free from dirt specks. Certain standard paper grades have been modified to perform on new scanning equipment.

In the printing-forms industry, presses have been designed to print continuous multiple ply forms by in-feeding several rolls simultaneously. This equipment requires that paper must have greater dimensional stability and must be manufactured to closer tolerances of weight and thickness.

Printing plants are being operated under controlled humidity and temperature to an increasing extent. The result is that greater attention must be given to the manufacture of paper that is in equilibrium with specified relative humidity.

Consumer

In any event, the common denominator is the consumer. Consumers are becoming increasingly sophisticated buyers, and, with this sophistication, has come the demand for concrete performance data. In order to provide this data, the manufacturer must test his product to obtain it. From a legal point of view, this test data may be invaluable in protecting the manufacturers and, as far as the consumer is concerned, he will be buying a more mature product as a result of it.

SUMMARY VIEWPOINT

Throughout this book we emphasized that most product evaluation problems involve a balance of costs for an optimum solution. Invariably, cost factors affected by a given decision follow different patterns, so that the best solutions never involve minimizing one cost factor at the expense of the others. This means that we do produce some defectives; we do have inefficient tests; we do have program delays; and we do fail to include some environments. If this were not true, we would be controlling some factors too tightly and not operating in an optimal fashion.

Remember that the techniques were presented in order to give initial direction to applicable areas. They were not intended to be definitive, except for a very rare circumstance. Their most important function was to increase understanding and to create awareness in the reader of approaches taken by others in related fields. An appreciation of the universality of the PT&E function, regardless of the industry, product, or service is of definite benefit in performing a better job wherever it may be or whatever your interest.

Finally, the broader our viewpoint, the more likely it is that we can avoid serious shortcomings in our PT&E programs. It is not valid to consider each evaluation problem in isolation. We have illustrated the benefits derived from an integrated systems approach to the test and evaluation of products and services.

As Issac Asimov prophetically states in the Foreword, "With improved product testing and greater care in manufacture, it may just be that, when our astronauts land on Mars with perfectly functioning retrorocket procedures, the switch on the dollar flashlight they bring along with them won't jam." We believe that with the application of the concepts and theories delineated in this book, the prophesy will be fulfilled, and we will be enlightened.

Appendix A
Standards and Calibration

Standards and calibration requirements are fundamental considerations of the test and evaluation function. No book would be complete without some recognition of this important consideration. The following block diagrams and flow charts suggest a method whereby the traceability of measurements accuracy is derived and maintained in a typical setup.

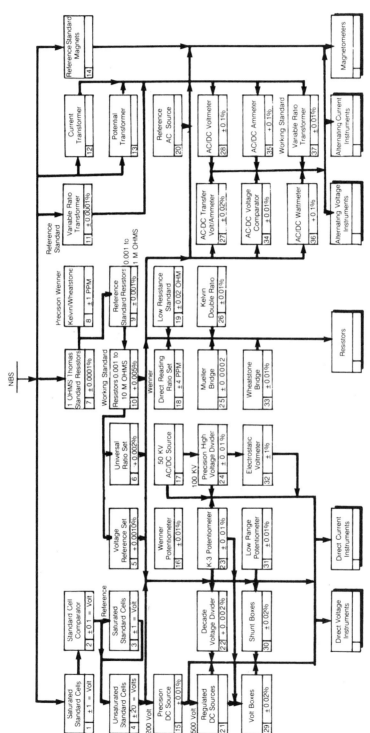

Figure A.1 Electrical measurements. This flow diagram depicts the direct relationship of electrical instrumentation measurements, to a particular standard submitted to the National Bureau of Standards. This diagram also indicates how the accuracy of NBS--certified instruments is traceable through the different levels of accuracy to a final calibration function performed on production equipment.

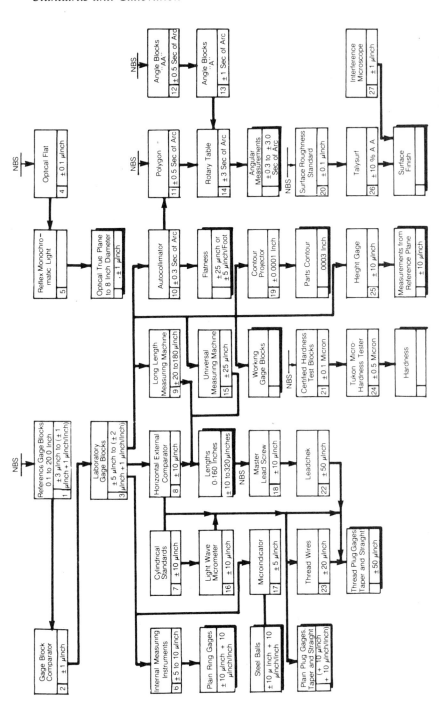

Figure A.2 Physical measurements. This flow diagram illustrates how the relationship between physical standards establishes an unbroken chain of traceability to the National Bureau of Standards.

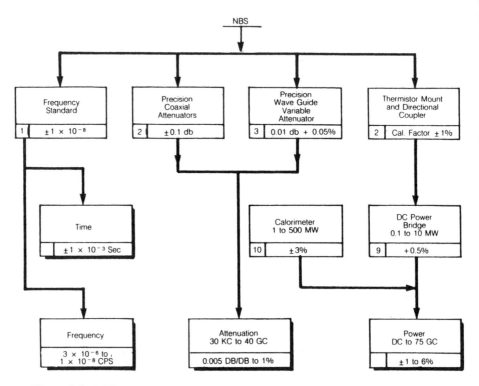

Figure A.3 RF/microwave measurements.

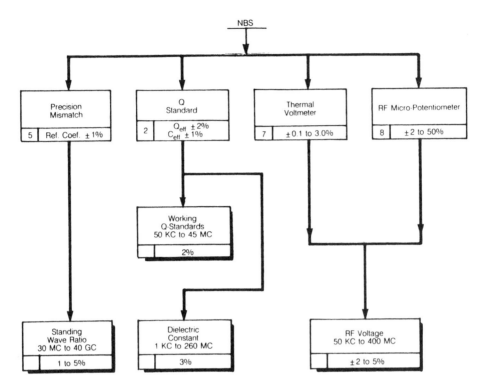

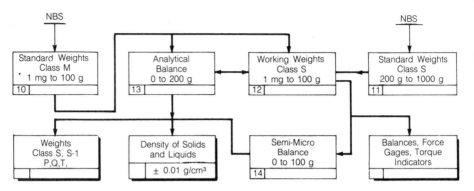

Figure A.4 Mass measurements.

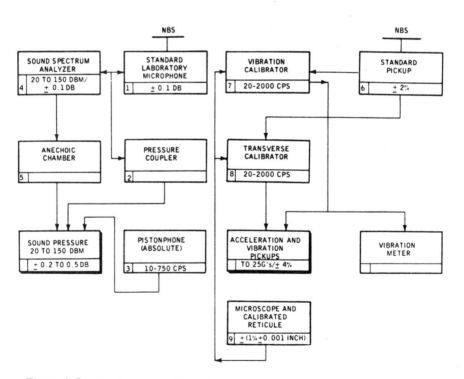

Figure A.5 Vibration, acoustic measurements.

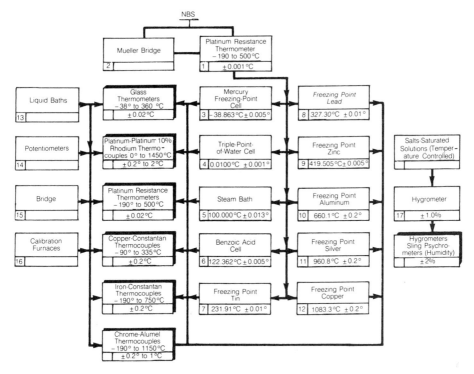

Figure A.6 Temperature and humidity measurements.

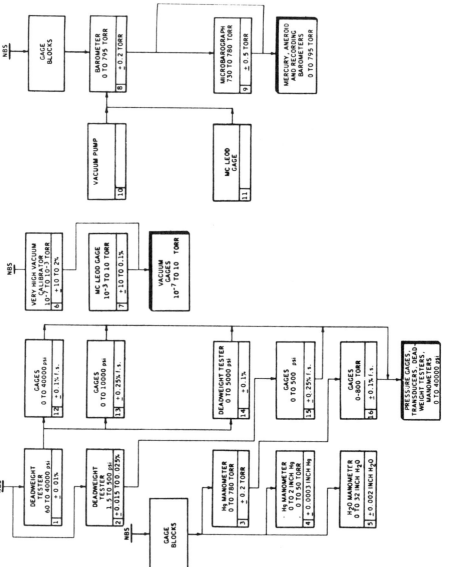

Figure A.7 Pressure/vacuum measurements

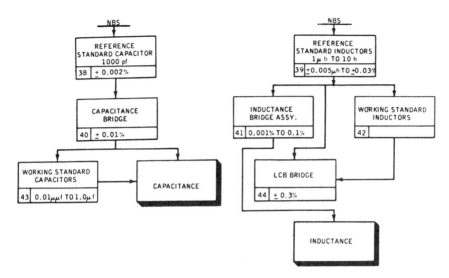

Figure A.8 Capacitance and inductance measurements.

Appendix B
Glossary

Acceleration Amplituded. The maximum zero-to-zero amplitude of acceleration of the fundamental frequency along a specified axis expressed in gravitational units (number of g).

Accessibility. A measure of the relative ease of admission to the various areas of an item.

Amplitude Linearity. The quality of having a constant "system gain" for all input amplitudes at any specific frequency.

Attribute. A qualitative characteristic (such as acceptable or rejectable, success or failure, rusted or not rusted, wet or dry, black or white, hit or miss) that can have two or more categories.

Attribute Data. Data denoting a qualitative characteristic. This type of data can have only discrete values and is derived by counting the number of times that each category occurs, such as four failures and six successes.

Availability. A measure of the degree to which an item is in the operable and commitable state of the start of the mission, when the mission is called for at an unknown (random) point in time.

Best Estimate. An estimator is said to give the "best estimate" of the true population parameter if it complies with the following requirements that are taken as the definition of the word "best": (1) the average of all possible values of the estimator equals the true population parameter; and (2) in any particular case the deviation of the estimator from the true population parameter is less than any other possible estimator.

Blockings.   In experimental design, a block is a homogeneous group of items, all treated under controlled conditions such as by the same operator, the same calibration of the measuring instrument, or the same short period of time.   The purpose of blocking is to reduce the effect of the heterogeneity of material and changing conditions by dividing the experiment into rational subdivisions.

Binomial Data.   Attribute data that have only two categories or only two possible outcomes such as success and failure.

Burn-In.   The operation of an item to stabilize its characteristics.

Capability.   A measure of the ability of an item to achieve mission objectives, given the conditions during the mission.

Checkout.   Tests or observations of an item to determine its conditions or status.

Complex Wave.   Any wave that is not sinusoidal.

Confidence Interval.   The range of values within which the true population parameter (mean or standard deviation) is expected to lie.   The confidence level associated with this interval is a probability statement expressing the proportion of the time in true value which is expected to be within the interval.

Confidence Level.   The confidence level is the probability of being right in our prediction or conclusions.   This value is equal to one minus the error of the first kind.   The magnitude of this error that can be tolerated should be established during the planning stage of the experiment (prior to data collecting), based on the consequences of being wrong, and thereby should establish the confidence level.

Confounding.   When certain comparisons can be made only for treatments in combination and not for separate treatments, the treatment effects in this case will be ambiguous.   Confounding is often a deliberate . feature of the experimental design but may arise from inadvertent imperfections.

Criterion.   The measurable characteristic used to evaluate the treatment effects.   Criteria can also be considered as the dependent variables used as a standard of reference to distinguish between the independent variable effects.   Velocity functioning time, voltage, rate of detonation, etc., can be criteria.

Critical Frequencies.   The fundamental frequencies at which the vibration is most likely to cause structural failure or malfunction of the test item.   The most critical frequencies need not always coincide with resonant frequencies.

Debugging.   A process to detect and remedy inadequacies, preferably prior to operational use.

Degrees of Freedom. The number of degrees of freedom is equal to the number of independent observations minus the number of parameters (such as the mean) estimated. That is, degrees of freedom usually equal the sample size minus one. In computing the variance, for example, only $(n-1)$ of the deviations from the mean can be independent. The nth deviation has to be restricted in order to make the sum of all n deviations total zero.

Demonstrated. That which has been proved by the use of concrete evidence gathered under specified conditions.

Dependability. A measure of the item operating condition at one or more points during the mission, including the effects of reliability, maintainability, and survivability, given the item condition(s) at the start of the mission. It may be stated as the probability that an item will (1) enter or occupy any one of the required operational modes during the specified mission, and (2) perform the functions associated with those operational modes.

Derating. (1) Using an item in such a way that applied stresses are below rates values, or (2) the lowering of the rating of an item in one stress field to allow an increase in rating of another stress field.

Displacement Amplitude (Double Displacement). The maximum peak-to-peak amplitude of displacement motion of the fundamental frequency along a specified axis.

Distortion. Any frequencies other than the fundamental frequency present on an acceleration input or response signal.

Downtime. See Time, Down.

Dry-Bulb Temperature Measurement. The measurement of the ambient temperature by a sensor.

Durability. See Reliability (of which this is a special case).

Effect. In statistics the meaning of the word effect is synonomous with the word difference. A treatment effect is the difference caused by the treatment, such as the difference in the measured results before and after the treatment.

Efficiency. An estimator or an experimental design is said to be efficient if a given precision can be obtained with a smaller sample size or with less time and cost.

Equalization. The process whereby the frequency response of a vibration system is adjusted to a flat or shaped spectrum.

Error. Chance variations are considered errors in statistics. Deviations from the expected value, due to chance, form the familiar bell-shaped normal curve. This is sometimes called the normal curve of error. Error, in the statistical sense, does not imply that a mistake has been made.

Error Mean Square. The error mean square is the variance and is also the square of the standard deviation. It is calculated by finding the sum of the squares of the deviations of the individual sample values from their mean and dividing by the number of degrees of freedom.

Error of Estimate. The difference between an estimated value and the true value.

Error of First Kind. If, as a result of a statistical test, the null hypothesis is rejected when it is true, then it is said that an error of the first kind is committed. This type of error is also called (1) the alpha error; (2) the producer's risk; and (3) the risk of rejecting good material. The magnitude of this error should be established from the consequences of being wrong and controlled at that level by calculating the required sample size.

Error of Observation. An error of observation arises from imperfections in the method of measurement or from human mistakes.

Error of Second Kind. If, as a result of a statistical test, the null hypothesis is accepted when it is false, then it is said that an error of the second kind is committed. This type of error is also called: (1) the beta error (2) the consumer's risk and (3) the risk of accepting poor material. After the error of the first kind has been proved, the error of the second kind is controlled by the sample size. This error is very important in ordnance work because it controls the probability of accepting poor material.

Estimate. An estimate is the particular value obtained by an estimator in a given set of circumstances.

Estimator. An estimator is the method of estimating a constant of a parent population. It is usually expressed as a function of sample values (such as the average) and therefore is a variable.

Experimental Error. Experimental error is the chance variation to be expected under controlled conditions. It is not the result of mistakes in experimental design or avoidable imperfections in technique.

Experimental Unit. An experimental unit is the smallest subdivision of the experimental material that can receive different treatments in the actual experiment. It is also known as a test specimen.

Factor. A factor is a quantity under examination (in an experiment) as a possible cause of variation. In practice the terms factor, treatment, and variable are loosely used interchangeably in this sense.

Factorial Experiment. An experiment that investigates all the possible treatment combinations that may be formed from the factor versions under investigation.

Failure. The inability of an item to perform within previously specified limits.

Failure Analysis. The logical, systematic examination of an item or its diagram(s) to identify and analyze the probability, causes, and consequences of potential and real failures.

Failure Dependent. One that is caused by the failure of an associated item(s). Not independent.

Failure Independent. One that occurs without being related to the failure of associated items. Not dependent.

Failure, Random. any failure whose occurrence is unpredictable in an absolute sense but which is predictable only in a probabilistic or statistical sense.

Failure Rate. The number of failures of an item per unit measure of life (cycles, time, miles, events, etc., as applicable for the item).

Faired Amplitude. The amplitude of the dotted line at any time during the pulse interval. This amplitude is used to eliminate unwanted frequencies of a recorded pulse to determine the basic shape of the input pulse.

Fall Time. The time required for the faired amplitude to fall from 90% to 10% of its maximum value.

Filtered Signal. Ideally a filtered signal has all distortion removed, leaving a pure sinusoidal signal. However, an input control signal should be considered a "filtered signal" if, in the fundamental frequency range of 50−500 cps, the signal has been passed through an 800 cps low-pass filter with at least an 18 dB/octave rolloff or, in the frequency range of 500−2000 cps, the signal has been passed through an 8000 cps low-pass filter with at least an 18 dB/octave rolloff. An output or response signal should be considered a filtered signal if distortion has been attenuated to such a degree that the amplitude measurement of the resulting signal agrees with the amplitude measurement of the pure fundamental within ±10%.

Fractional Factorial Experiment. This is a fractional part of a factorial experiment. When three or more factors are used in a factorial experiment only a fractional part ($1/2$, $1/4$, $1/8$) of the total number of possible combinations need to be used if certain of of the interactions can be considered negligible. This device can be resorted to without loss of efficiency when the number of factors to be investigated makes the full factorial so large that it is impractical to use.

Frequency Response. The "system gain" for a constant input amplitude as a function of frequency.

Fundamental Frequency. The particular forcing frequency of the vibration machine power supply.

Human Engineering. The area of human factors that applies scientific knowledge to the design of items to achieve effective man-machine integration and utilization.

Human Factors.   A body of scientific facts about human characteristics. The term covers all biomedical and psychosocial considerations: it includes, but is not limited to, principles and applications in the areas of human engineering, personnel selection, training, life support, job performance aids, and human performance evaluation.

Human Function.   The function allocated to the human component of a system.

Human Performance.   A measure of human functions and actions in a specified environment.

Hypothesis.   A hypothesis is a contention based on preliminary observation of what appears to be fact.   It is the prediction derived from post experience that is to be verified or rejected by experimentation. Natural "laws" are hypotheses that have been subjected to various tests and have been accepted.   In statistical tests, two hypotheses are used: (1) The *null hypothesis* is a hypothesis of "no difference." This is the assumption that the contemplated changes will make no difference. This hypothesis is formulated for the express purpose of being rejected in the process of controlling the error of the first kind.   (2) The *alternative hypothesis* is the operational statement of the experimenter's prediction.   It is the positive statement that the changes will make a detectable difference.   If the resultant data reject the null hypothesis the alternative hypothesis will be accepted.

Independence.   Measurements are independent if the taking of one does not affect any of the others.   That is, there is no correlation among them.   Treatment effects are said to be independent if, in an orthogonal experiment, there is no interaction.

Inherent.   Achievable under ideal conditions, generally derived by analysis, and potentially present in the design.

Input Control Signal.   The signal from the input control transducer.

Interaction.   The interaction is a measure of the extent to which the effect of changing the level of one factor depends on the level of another factor.   Interaction is said to be present when a certain particular combination of treatments produces unusual (unpredictable) results. Only factorial-type experiments can measure interaction effects.

Inventory, Active.   That group of items assigned an operational status.

Inventory, Inactive.   That group of items being held in reserve for possible future commitment to the operational inventory.

Item, Interchangeable.   An item that (1) possesses such functional and physical characteristics as to be equivalent in performance, reliability, and maintainability, to another item of similar or identical purpose; and (2) is capable of being exchanged for the other item (a) without selection for fit or performance, and (b) without alteration of the items themselves or of adjoining items, except for adjustment.

Item, Replaceable. One which is interchangeable with another item, but which differs physically from the original item in that the installation of the replaceable item requires operations such as drilling, reaming, cutting, filing, shimming, etc., in addition to the normal application and methods of attachment.

Item, Substitute. One which possesses such functional and physical characteristics as to be capable of being exchanged for another only under specified conditions or for particular applications and without alteration of the items themselves or of adjoining items.

Levels. The level of a factor (or treatment) denotes the intensity with which it is used or applied. Levels of a factor may be either qualitative (such as presence and absence of the treatment) or quantitative (such as the number of volts applied).

Life Support. The area of human factors that applies scientific knowledge to items that require special attention or provisions for health promotion, biomedical aspects of safety, protection, sustenance, escape, survival, and recovery of personnel.

Main Effects. A main effect is the average difference(s) between (or among) the levels of a variable or treatment when averaged over all the other treatments which form a part of the same orthogonal experiment. If significant interaction effects are present, care must be taken in stating the main effects. In such cases the level of the interacting treatment associated with the stated main effect must also be stated.

Maintainability. A characteristic of design and installation which is expressed as the probability that an item will be retained in or restored to a specified condition within a given period of time, when the maintenance is performed in accordance with prescribed procedures and resources.

Maintenance. All actions necessary for retaining an item in or restoring it to a specified condition.

Maintenance, Corrective. The actions performed, as a result of failure, to restore an item to a specified condition.

Maintenance, Preventive. The actions performed in an attempt to retain an item in a specified condition by providing systematic inspection, detection, and prevention of incipient failure.

Man-Function. See Human Function.

Maximum Amplitude. The maximum attained height of the shock pulse, measured from the zero line.

Maximum Faired Amplitude. The maximum height of the faired amplitude, measured from the zero line. This value is used in computing the rise time, fall time, and pulse duration, and is the amplitude specified in detailed test specifications.

Mean Maintenance Time.   The total preventive and corrective maintenance time divided by the total number of preventive and corrective
maintenance actions during a specified period of time.

Mean Time Between Failures (MTBF).   For a particular interval, the
total functioning life of a population of an item divided by the total
number of failures within the population during the measurement interval.   The definition holds for time, cycles, miles, events, or other measures of life units.

Mean Time Between Maintenance.   The mean of the distribution of the
time intervals between maintenance actions (either preventive, corrective, or both).

Mean Time to Repair (MTTR).   The total corrective maintenance time
divided by the total number of corrective maintenance actions during a
given period of time.

Measurement Limits (when measuring system is specified).   When the
method and/or accuracy of measurement or the measurement equipment
is specified by the product requirements, measurement uncertainty has
been adequately allowed for in establishing the measurement limits.
Specified measurement equipment may include universal measuring
machines and is not limited to gage designed for particular parts.   A
part will be considered conforming to design intent if the measurement
is within the specified limits.   The absolute tolerance requirement does
not apply under these conditions.   In comparing measured (observed)
values with the stated tolerances, it is not permissible to round the digit
in the decimal place one unit beyond the last place in the stated tolerance.

Mission.   The objective or task, together with the purpose, which
clearly indicates the action to be taken.

Multiple Filter Equalizer.   Any equalizer that equalizes by the process
of adjusting the gain of filters having fixed-center frequencies and bandwidths with a random noise input to the vibration table.

Normal Distribution.   The physical appearance of a normal distribution
is the familiar bell-shaped curve.   A normal distribution cannot be represented by only a single curve.   It is actually a family of curves whose
areas under them are distributed in a very specific manner.   A normal
curve has the following properties: (1) continuous (2) symmetrical (3)
unimodal (4) asymptotic to x axis (5) completely described by the mean
and standard deviation (6) the distance between the ordinate of the
mean and the inflation point on either half of the curve is equal to the
standard deviation (7) the area included between the ordinates drawn
through the two inflection points equals 68.27% of the total area under
the curve.

Operable. The state of being able to perform the intended function.

Operational. Of, or pertaining to, the state of actual usage.

Output on Response Signals. The signals from transducers mounted on or monitoring the motion of points on the test item.

Overall Test Level. The measurement of rms acceleration at the input accelerometer with a filter in the circuit limiting the measurement bandwidth to the test bandwidth.

Parameter. A parameter is a quantity such as the mean or standard deviation, calculated from a population. The population mean and standard deviation are parameters and, as such, are constants. In actual practice, parameters are usually unknown.

Peak-Notch Equalizer. Any equalizer that equalizes by the process of using filters of adjustable gain and bandwidth with a sinusoidal input applied to the vibration table.

Personnel Subsystem. A management concept which considers that functional part of a system which provides, through effective development and implementation of its various elements, the specified human performance necessary in the operation, maintenance, support, and control of the system in a specified environment.

Point Estimate. This is one of the two principal bases of estimation in statistical analysis. Point estimation endeavors to give the best single estimated value of a parameter, as compared with interval estimation which specifies a range of values. Since a point estimate includes an error of measurement, the difference between a point and an interval is not always clear. In interpretation they often amount to the same thing.

Population. A population is any set of individuals or objects having some common observable characteristic. The term population may refer either to the individuals measured or to the measurements themselves. A population is usually considered to consist of an *infinite* number of individuals. The curve of the normal distribution graphically represents a population.

Power Spectral Density. The square of the rms acceleration at a specified frequency, divided by the bandwidth used to measure it (g^2/cps), as the bandwidth approaches zero.

Predicted. That which is expected at some future date, postulated on analysis of past experience.

Precision. Precision is a property of the measuring system and refers to the ability of the system to reproduce previous results. Precision should be distinguished from accuracy, which refers to the magnitude of the difference between the observed values and the true value of the characteristic being measured. Precision also should be distinguished from the sensitivity of the measuring system, which is the ability of the system to detect actual variations that occur. An insensitive system will give the false impression of high precision (small variation).

Primary Reference Standard. The prime reference of the standard and calibration system for a particular type of measurements. Periodic calibration by the National Bureau of Standards assures its consistency with legal standards. The use of the term "primary" is relative within the system and does not, in any sense, imply independence of the national legal standards.

Primary Shock Pulse. The portion of the generated acceleration-time history that represents the pulse specified in the detail test specification. This portion of the history is the most important in shock testing.

Probability. In applied statistics, probability can be considered as a relative frequency or a simple proportion. Probability is the relative frequency of events in a very long sequence of trials. For example, the probability of a particular coin falling heads up is the ratio of the number of heads occurring to the total number of trials in a sequence of trials. In somewhat similar fashion a normal distribution can be formed from a very large body of data. As a result, the area under the normal curve is used as a measure of probability.

Pulse Duration. The time when the faired amplitude is above 10% of its maximum value.

Rating. The value of an item parameter which can be attained under specified conditions.

Randomization. The word randomization has a very special technical meaning in statistics. It means rearranging a group of items or numbers into a series or sequence having no recognizable pattern. The essential feature of randomization is that it should be an objective impersonal procedure. Whether proper randomization has been obtained should not be determined by examining the individuals randomized but, instead, by examining the properties of the procedure by which randomization was accomplished. The objectives of randomizing are as follows: (1) to give the laws of chance free play (2) to give every possible sequence an equally likely chance of occurring (3) to assure that adjacent individuals are completely independent (4) to remove biases of any kind (5) to remove systematic error.

Random Noise. A fluctuating quantity (such as sound pressure or an electrical wave) whose instantaneous amplitudes occur as a function of time according to a normal (Gaussian) distribution.

Random Vibration Test. A vibration test using random noise as the input signal.

Ready Rate, Operational (Combat). Percent of assigned items capable of performing the mission of function for which they were designed, at a random point in time.

Redundancy. The existence of more than one means for accomplishing a given function. Each means of accomplishing the function need not necessarily be identical.

Redundancy, Active. That redundancy wherein all redundant items are operating simultaneously instead of being switched on when needed.

Redundancy, Standby. That redundancy wherein the alternative means of performing the function is inoperative until needed and is switched on upon failure of the primary means of performing the function.

Reference Point. The term used to periodic control instrument checks and temperature distribution checks as the location of the reference sensor (or standard sensor). It should be the geometric center of that volume which has met the requirements.

Relative Humidity. The ratio of the quantity of water vapor present in the atmosphere to the quantity which would saturate the atmosphere at the existing temperature.

Reliability. The probability that an item will perform its intended function for a specified interval under stated conditions.

Reliability, Human Performance. The probability that man will accomplish all required human functions under specified conditions.

Repair. See Maintenance Corrective.

Replication. Replication is the performance of an experiment in its entirety, one or more times. The purpose of two or more replications is usually to obtain an independent measure of the sampling or experimental error. Replication should be distinguished from repetition in that replication means repetition carried out under the same conditions, at the same time, by the same operators, with the same instruments, and with the same homogeneous material. A replication is sometimes considered a block.

Resonance. A mode of resonance exists when an increase or decrease in the frequency of excitation causes a decrease in the fundamental response of the test item.

Rise Time. The time required for the faired amplitude to rise from 10 to 90% of its maximum value.

Safety. The conservation of human life and its effectiveness, and the prevention of damage to items, consistent with mission requirements.

Sample. Any finite subset of a population is a sample of that population.

Sample Size. The sample size is the number of items or individual values in the sample.

Secondary Reference Standard. The standard specified for use at a secondary standards laboratories for the calibration of secondary transfer standards (and/or instruments, gages, and testers in special cases).

Serviceability. A measure of the degree to which servicing of an item will be accomplished within a given time under specified conditions.

Servicing. The replenishment of consumables needed to keep an item in operating condition, but not included any other preventive maintenance or any corrective maintenance.

Sigma. The level of random noise signal that is equal to the rms level of the random signal. "Peaks of 3-sigma" is a term meaning that the amplitude has individual peaks equal to three times the rms value.

Stabilized. The term used in temperature distribution measurements to define the interior empty chamber condition that exists at a test temperature when the temperature measured at a reference point does not vary more than $\pm 2.0°F$ $(1.11°C)$ during a ten-minute interval.

Standards. The physical embodiment of a defined unit of measure under specified conditions.

Standard Deviation. (a) *Definition.* The standard deviation is a measure of the variation among the individual values in a sample and a measure of the dispersion among the individual values in a frequency distribution. It is the most efficient measure of precision and is designated by the lower case letter "s". This value is large for large variation (poor precision) and small for small variations (good precision). Although the word "error" is sometimes used in referring to the standard deviation or its square (the variance), these values can measure only precision in the true sense of the word. They do not measure accuracy.

If the term "standard deviation" is stated alone and is not modified or otherwise qualified by an accompanying word or phrase, it is understood that the term refers to the standard deviation of the individual sample measurements. This value can be calculated from the sample data and is *variable.*

There are two additional kinds of standard deviations:

1. The *population standard deviation* which is a constant and cannot be calculated from the sample data. This value, designated by the small Greek letter sigma, is usually considered unknown unless a very large body of data is collected to measure it or unless it is assigned a value as in a specification requirement.

2. The *standard deviation of the mean* is a measure of the variation among several sample averages. This value can be calculated from sample data and it is a variable. It is usually designated by the lower case letter "s" with the subscript $\bar{X}$. If all the sample sizes are equal, this

value can be calculated by dividing the standard deviation of the individual sample values by the square root of the number of individual values in each of the samples.

  (b) *Calculation of the standard deviation for variable type data:*

$$s = \frac{\sqrt{-\sum\limits_{i=1}^{i=n}(\bar{X}x_i)^2}}{n-1} = \sqrt{\frac{\sum\limits_{i=1}^{i=n}(x_i)^2 - \dfrac{\left(\sum\limits_{i=1}^{i=n}x_i\right)}{n}}{n-1}}$$

where

s	= Sample standard deviation of the individual values.
$\sum\limits_{i=1}^{i=n}$	= This symbol means to add all of the n quantities designed by the parenthesis. It is read: sum from $i = 1$ to $i = n$.
$\bar{X}$	= Sample average.
x_i	= Any one of the n values that make up the sample.
$(n-1)$	= Number of degrees of freedom associated with the standard deviation.
s^2	= Sample variance of the individual values.
$s/\sqrt{n}$	= Sample standard deviation of the mean $(\bar{s}_x)$.

Statistic. A statistic is a summary value calculated from a sample of values. The sample mean is a statistic and, as such, is a variable, not a constant.

Statistics. The subject of statistics is the science of collecting, analyzing, and interpreting numerical data.

Statistical Significance. A difference or an effect is said to be statistically significant if it is greater than that expected as the result of chance alone. If the probability (chance) is very small that a value came from a particular population, the difference between that value and the mean of the population is said to be statistically significant.

Storage Life (Shelf Life). The length of time that an item can be stored under specified conditions and still meet specified requirements.

Survivability. The measure of the degree to which an item will withstand hostile man-made environment and not suffer abortive impairment of its ability to accomplish its designated mission.

System Effectiveness. A measure of the degree to which an item can be expected to achieve a set of specific mission requirements, and which may be expressed as a function of availability, dependability, and capability.

Temperature Effectiveness. A measure of the degree to which an item can be expected to achieve a set of specific mission requirements, and which may be expressed as a function of availability, dependability, and capability.

Temperature Differential (Wet-Bulb Depression). The difference between dry-bulb temperature and the wet-bulb temperature.

Time, Active. That time during which an item is in the operational inventory.

Time, Adjustment or Calibration. That element of maintenance time during which the needed adjustments of calibrations are made.

Time, Administrative. Those elements of delay time that are not included in supply delay time.

Time, Alert. That element of uptime during which an item is thought to be in specified operating condition and is awaiting a command to perform its intended mission.

Time, Checkout. That element of maintenance time during which performance of an item is verified to be in specified condition.

Time, Cleanup. That element of maintenance time during which the item is enclosed and extraneous material not required for operation is removed.

Time, Delay. That element of downtime during which no maintenance is being accomplished on the item because of either supply delay or administrative reasons.

Time, Down (Downtime). That element of time during which the item is not in condition to perform its intended function.

Time, Fault Correction. That element of maintenance time during which a failure is corrected by (1) repairing in place (2) removing, repairing, and replacing or (3) removing and replacing with a like serviceable item.

Time, Fault Location. That element of maintenance time during which testing and analysis is performed on an item to isolate a failure.

Time, Inactive. That time during which an item is in reserve (in the inactive inventory).

Time, Item Obtainment. That element of maintenance time during which the needed item or items are being obtained from designated organizational stockrooms.

Time, Mission. That element of uptime during which the item is performing its designated mission.

Time, Modification. The time necessary to introduce any specific change(s) to an item to improve its characteristics or to add new ones.

Time, Preparation. That element of maintenance time needed to obtain the necessary test equipment and maintenance manuals, and set up the necessary equipment to initiate fault location.

Time, Reaction. That element of uptime needed to initiate a mission, measured from the time the command is received.

Time, Supply Delay. That element of delay time during which a needed item is being obtained from other than the designated organizational stockrooms.

Time, Turn-Around. That element of maintenance time needed to service or checkout an item for recommitment.

Time, Up (Uptime). That element of active time during which an item is either alert, reacting, or performing a mission.

Tolerance, Absolute Tolerances. Tolerances expressed on drawings and specifications are absolute values. The term "absolute" means that the true values of characteristics, not merely the measured values, must be within the tolerances stated. The product must conform to absolute tolerance to an acceptable degree. To judge the degree of conformance, the inherent measurement uncertainty in relation to the specified tolerances must be considered. Information concerning production process distribution (both centering and dispersion), if available, may also be used in determining conformance.

Transfer Standards (primary and secondary). Standards used to transfer values from standards of a given level of accuracy to other standards one step lower in accuracy, or to gauges, instruments, or testers.

Transmissibility. A ratio of the transmitted force to the disturbing force, or the ratio of the amplitudes of the fundamental response acceleration to the fundamental input acceleration.

Treatment. In experimentation, a treatment is a stimulus that is applied in order to observe the effect on the experimental situation. A treatment may refer to a physical substance, a procedure, or anything that is capable of controlled application. In statistical parlance a treatment is the variable being studied or the experimental condition.

True Value. The true value is another expression for a population parameter such as the population mean or standard deviation. The true value can also be the expected value or the theoretical value.

Uptime Ratio. The quotient of uptime divided by uptime plus downtime.

Validation. A procedure which provides, by reference to independent sources, evidence that an inquiry is free from bias, or otherwise conforms to its declared purpose. In statistics it is usually applied to a sample investigation to show that the sample is reasonably representative of the population and that the information collected is accurate.

Variable. Any quantity of measurable characteristic that varies. More precisely, in statistics a variable is any quantity that can have any one of a specified set of values.

Variable Data. A term used to describe a type of data that can vary on a continuous scale from zero to infinity. Weight in pounds, length in feet, EMF in volts, and temperature in degrees are variable types of data.

Variance.   A measure of variation in a sample, or dispersion in a frequency distribution. The variance is equal to the square of the standard deviation.

Velocity Change (ΔV).   The change in velocity from the start of the primary pulse to the end of the pulse.

Wearout.   The process of attrition that results in an increase of the failure rate with increasing age (cycles, time, miles, events, etc., as applicable for the item).

Wet-Bulb Temperature Measurement.   The measurement of the equilibrium temperature caused by water evaporation in moving air.

White Gaussian Noise.   White Gaussian noise is random noise with equal energy per unit frequency bandwidth (constant g^2/cps).

White Noise.   Noise having equal energy per unit frequency bandwidths over a specified total frequency band (constant g^2/cps). White noise need not be random.

Zero Line.   The line representing the zero acceleration condition of the shock machine carriage when it is at rest or moving with constant velocity.

Index